WEATHER RADAR TECHNOLOGY
BEYOND NEXRAD

Committee on Weather Radar Technology Beyond NEXRAD

Board on Atmospheric Sciences and Climate

Division on Earth and Life Studies

National Research Council

NATIONAL ACADEMY PRESS
Washington, D.C.

NATIONAL ACADEMY PRESS • 2101 Constitution Avenue, N.W. • Washington, DC 20418

NOTICE: The project that is the subject of this report was approved by the Governing Board of the National Research Council, whose members are drawn from the councils of the National Academy of Sciences, the National Academy of Engineering, and the Institute of Medicine. The members of the committee responsible for the report were chosen for their special competences and with regard for appropriate balance.

This study was supported by Contract No. 56-DKNA-1-95101 between the National Academy of Sciences and the National Oceanic and Atmospheric Administration (NOAA), Contract No. DTFA0101G10185 between the National Academy of Sciences and the Federal Aviation Administration, and Grant No. N00014-00-1-0912 between the National Academy of Sciences and the Office of Naval Research. Additional funding was provided by the U.S. Air Force through the NOAA contract. The views and any opinions, findings, conclusions, or recommendations expressed in this publication are those of the author(s) and do not necessarily reflect the views of the organizations or agencies that provided support for the project.

International Standard Book Number 0-309-08466-0

Additional copies of this report are available from National Academy Press, 2101 Constitution Avenue, N.W., Lockbox 285, Washington, DC 20055; (800) 624-6242 or (202) 334-3313 (in the Washington metropolitan area); Internet, http://www.nap.edu

Printed in the United States of America

THE NATIONAL ACADEMIES

National Academy of Sciences
National Academy of Engineering
Institute of Medicine
National Research Council

The **National Academy of Sciences** is a private, nonprofit, self-perpetuating society of distinguished scholars engaged in scientific and engineering research, dedicated to the furtherance of science and technology and to their use for the general welfare. Upon the authority of the charter granted to it by the Congress in 1863, the Academy has a mandate that requires it to advise the federal government on scientific and technical matters. Dr. Bruce M. Alberts is president of the National Academy of Sciences.

The **National Academy of Engineering** was established in 1964, under the charter of the National Academy of Sciences, as a parallel organization of outstanding engineers. It is autonomous in its administration and in the selection of its members, sharing with the National Academy of Sciences the responsibility for advising the federal government. The National Academy of Engineering also sponsors engineering programs aimed at meeting national needs, encourages education and research, and recognizes the superior achievements of engineers. Dr. William A. Wulf is president of the National Academy of Engineering.

The **Institute of Medicine** was established in 1970 by the National Academy of Sciences to secure the services of eminent members of appropriate professions in the examination of policy matters pertaining to the health of the public. The Institute acts under the responsibility given to the National Academy of Sciences by its congressional charter to be an adviser to the federal government and, upon its own initiative, to identify issues of medical care, research, and education. Dr. Harvey V. Fineberg is president of the Institute of Medicine.

The **National Research Council** was organized by the National Academy of Sciences in 1916 to associate the broad community of science and technology with the Academy's purposes of furthering knowledge and advising the federal government. Functioning in accordance with general policies determined by the Academy, the Council has become the principal operating agency of both the National Academy of Sciences and the National Academy of Engineering in providing services to the government, the public, and the scientific and engineering communities. The Council is administered jointly by both Academies and the Institute of Medicine. Dr. Bruce M. Alberts and Dr. William A. Wulf are chair and vice-chair, respectively, of the National Research Council.

COMMITTEE ON WEATHER RADAR TECHNOLOGY BEYOND NEXRAD

PAUL L. SMITH (*chair*), South Dakota School of Mines and Technology, Rapid City
DAVID ATLAS, Consultant, Bethesda, Maryland
HOWARD B. BLUESTEIN, University of Oklahoma, Norman
V. CHANDRASEKAR, Colorado State University, Fort Collins
EUGENIA KALNAY, University of Maryland, College Park
R. JEFFREY KEELER, National Center for Atmospheric Research, Boulder, Colorado
JOHN McCARTHY, Naval Research Laboratory, Monterey, California
STEVEN A. RUTLEDGE, Colorado State University, Fort Collins
THOMAS A. SELIGA, Volpe National Transportation Systems Center, Cambridge, Massachusetts
ROBERT J. SERAFIN, National Center for Atmospheric Research, Boulder, Colorado
F. WESLEY WILSON, JR., National Center for Atmospheric Research, Boulder, Colorado

NATIONAL RESEARCH COUNCIL STAFF

VAUGHAN C. TUREKIAN, Study Director
DIANE L. GUSTAFSON, Administrative Associate
ROB GREENWAY, Project Assistant
ELIZABETH A. GALINIS, Project Assistant

Preface

Weather radar is a vital instrument for observing the atmosphere to help provide weather forecasts and issue weather warnings to the public. The current Next Generation Weather Radar (NEXRAD) system provides Doppler radar coverage to most regions of the United States (NRC, 1995). This network was designed in the mid 1980s and deployed in the 1990s as part of the National Weather Service (NWS) modernization (NRC, 1999). Since the initial design phase of the NEXRAD program, considerable advances have been made in radar technologies and in the use of weather radar for monitoring and prediction. The development of new technologies provides the motivation for appraising the status of the current weather radar system and identifying the most promising approaches for the development of its eventual replacement.

The charge to the committee was:

To determine the state of knowledge regarding ground-based weather surveillance radar technology and identify the most promising approaches for the design of the replacement for the present Doppler Weather Radar. Specifically, the committee will:

1. *Examine the state of the present radar technologies;*
2. *Identify new processes for data analyses; and*
3. *Estimate the maturity of the various capabilities and identify the most promising approaches.*

The committee included experts in radar technologies, meteorological applications, computer-processing capabilities for data handling, and application to numerical models.

To perform the charge, the committee held three information-gathering meetings. During the first meeting in April 2001, the sponsoring agencies [National Oceanic and Atmospheric Administration (NOAA), Federal Aviation Administration (FAA), U.S. Air Force (USAF), and U.S. Navy] provided briefings on their weather radar related activities and potential future needs. During the second and third meetings (September 2001 and November 2001), experts in radar design and application briefed the committee on current and anticipated developments.

This report presents a first look at potential approaches for future upgrades to or replacements of the current weather radar system. The need, and schedule, for replacing the current system has not been established, but the committee used the briefings and deliberations to assess how the current system satisfies the current and emerging needs of the operational and research communities and identified potential system upgrades for providing improved weather forecasts and warnings. The time scale for any total replacement of the system (20- to 30-year time horizon) precluded detailed investigation of the designs and cost structures associated with any new weather radar system. The committee instead noted technologies that could provide improvements over the capabilities of the evolving NEXRAD system and recommends more detailed investigation and evaluation of several of these technologies. In the course of its deliberations, the committee developed a sense that the processes by which the eventual replacement radar system is developed and deployed could be as significant as the specific technologies adopted. Consequently, some of the committee's recommendations deal with such procedural issues.

The report is divided into seven chapters. Chapter 1 notes the role of radar as one important part of the broader weather and climate observing and predicting system. Chapter 2 presents a brief overview of the current, but evolving, NEXRAD system and describes some of the shortcomings that advanced radar and supporting technologies might help to overcome. Chapter 3 reviews those advanced technologies that appear to offer promising opportunities for improving upon the capabilities possessed by the NEXRAD system. Chapter 4 describes variety of network configurations and novel platforms that might be part of a future radar observing system. Then Chapter 5 considers ways in which the improved capabilities of the next generation radar system would enhance the products used to support the primary functions of weather observing and forecasting. Recommendations developed from the earlier discussions are summarized in Chapter 6, and Chapter 7 presents some concluding remarks.

Because the subject of this report is radar technology, much of the text (especially chapter 3) uses highly technical terminology. Readers unfamiliar with this terminology may consult IEEE (1990), Barton et al. (1991), Doviak and

Zrnic (1993), Bringi and Chandrasekar (2001), or Skolnik (2001) for definitions and explanations.

The findings in this report encompass a broad range of scientific and technological capabilities, some of which are assuredly within reach in the near term, and others of which are visionary. The feasibility of the more visionary capabilities depends upon a variety of factors such as the evolution of enabling technologies and advances in basic understanding. Although one cannot anticipate the specific advances that will emerge, rapid progress can be expected as a result of the stepwise gains in scientific insight from the application of new technology and the feedback of that insight to further advances in technology. Moreover, further developments will depend upon the evolution of the political, social, and economic environment in the nation and the world. The technologies recommended in this report have the potential to mitigate some of the limitations of the NEXRAD system, but questions about technical feasibility remain, and benefit-cost analyses will be required to identify those approaches most suitable for the design of the future weather surveillance radar system. As a result the committee does not prioritize the recommendations, though it has grouped the recommendations into categories (Chapter 6) to facilitate any future prioritization.

The committee wishes to acknowledge the assistance of those experts who helped the committee with its assessment of the promising directions for developing enhanced capabilities in the next generation weather radar system by providing information about evolving radar technologies and evolving applications of weather radar data: James Belville, Rit Carbone, Russell Cook, Tim Crum, Dustin Evancho, Stephen Del Greco, Jim Evans, John Garnham, Jamie Hawkins, Sheldon Katz, Jeff Kimpel, Witold Krajewski, Ed Mahoney, Dave McLaughlin, Peter Meischner, Robert Saffle, Charles Schilling, Merrill Skolnik, Dan Strawbridge, Mark Surmeir, Jim Wilson, and Dusan Zrnic. In addition, the committee expresses appreciation to Vaughan Turekian, study director, and to Carter Ford, Diane Gustafson, Elizabeth Galinis, and Rob Greenway for their able and energetic assistance in organizing and supporting the activities of the committee during the preparation of this report.

Paul L. Smith
Chair
Committee on Weather Radar
Technology Beyond NEXRAD

Acknowledgments

This report has been reviewed in draft form by individuals chosen for their diverse perspectives and technical expertise, in accordance with procedures approved by the National Research Council's Report Review Committee. The purpose of this independent review is to provide candid and critical comments that will assist the institution in making its published report as sound as possible and to ensure that the report meets institutional standards for objectivity, evidence, and responsiveness to the study charge. The review comments and draft manuscript remain confidential to protect the integrity of the deliberative process. We wish to thank the following individuals for their review of this report:

RICHARD E. CARBONE, National Center for Atmospheric Research, Boulder, Colorado
STEVEN CLIFFORD, University of Colorado, Boulder, Colorado
ANDREW CROOK, National Center for Atmospheric Research, Boulder, Colorado
WITOLD F. KRAJEWSKI, The University of Iowa, Iowa City, Iowa
LESLIE R. LEMON, Basic Commerce and Industries, Inc., Independence, Missouri
MARGARET A. LEMONE, National Center for Atmospheric Research, Boulder, Colorado
ANDREW L. PAZMANY, University of Massachusetts, Amherst, Massachusetts

Although the reviewers listed above have provided constructive comments and suggestions, they were not asked to endorse the report's conclusions or recommendations, nor did they see the final draft of the report before its release.

The review of this report was overseen by Douglas Lilly, University of Oklahoma. Appointed by the National Research Council, he was responsible for making certain that an independent examination of this report was carried out in accordance with institutional procedures and that all review comments were carefully considered. Responsibility for the final content of this report rests entirely with the authoring committee and the institution.

Contents

Summary

Weather radar furnishes essential observations of the atmosphere used in providing weather forecasts and issuing weather warnings to the public. The primary weather surveillance radar system operated by U.S. agencies is the WSR-88D (NEXRAD) system, which consists of about 150 nearly identical radars deployed over the United States and some overseas locations in the 1990s. Data from this system support activities of the National Weather Service (NWS), Federal Aviation Administration (FAA), and Department of Defense (DoD). The data are also distributed to a wide variety of other users, including private sector organizations and the media.

Since the design of the NEXRAD system, important new radar technologies and methods for designing and operating radar systems have been developed. These advances provide the motivation for appraising the status of the current weather radar system and identifying the most promising approaches for the development of its eventual replacement. In order to address this issue, a National Research Council (NRC) committee was convened, charged with the following task:

To determine the state of knowledge regarding ground-based weather surveillance radar technology and identify the most promising approaches for the design of the replacement for the present Doppler Weather Radar. Specifically, the committee will:

1. *Examine the state of the present radar technologies;*
2. *Identify new processes for data analyses; and*
3. *Estimate the maturity of the various capabilities and identify the most promising approaches.*

1

The committee included experts in radar technologies, meteorological applications, computer-processing capabilities for data handling, and application to numerical models.

In the summary each of the committee's recommendations appears under the section of the report in which it is introduced. The recommendations in boldface italics deal with technologies that are deemed worthy of consideration in the development of the future replacement for the current NEXRAD system. They are categorized as "near-term," "far-term," or "visionary."[1] The committee also felt that the processes by which the future system is developed and deployed could be as significant as the technologies. The recommendations in standard italics refer to such procedural issues, and have no assigned priority.

The feasibility of the "far-term" and "visionary" technologies depends upon a variety of factors such as the evolution of enabling technologies and advances in basic understanding. Moreover, further developments will depend upon the evolution of the political, social, and economic environment in the nation and the world. In-depth feasibility studies will be required to determine which approaches are most likely to provide the needed improvements. The committee encourages the agencies that commissioned this study to follow through with the investigations necessary to establish the technical feasibility of the "far-term" and "visionary" technologies and to conduct benefit-cost analyses of the feasible ones.

RADAR IN THE ATMOSPHERIC OBSERVING AND PREDICTING SYSTEMS

Weather forecasting and warning applications are relying increasingly on integrated observations from a variety of systems that are asynchronous in time and are non uniformly spaced geographically. Weather radar is a key instrument that provides rapid update and full volumetric coverage. On regional scales, the combination of the primary radar with subsidiary radars (either fixed or mobile) with satellite data, with automated meteorological measurements from aircraft, and with a network of ground-based meteorological instruments reporting in real time has been shown to provide enhanced nowcasting and short-term forecasting capabilities. Such capabilities improve severe local storm warnings (including forecasts of storm initiation, evolution, and decay), and they support activities such as construction, road travel, the needs of the aviation system (both civil and military), and recreation.

---

[1]The committee uses the term "near term" for those technologies for which the capabilities exist currently and could be implemented even before the development of the replacement NEXRAD. "Far-term" technologies are those that could be available within the time period covered by this report (25–30 years), though they will require continued scientific and technological development before they could be implemented. "Visionary" technologies are those that may or may not be ready for operational use within the 25- to 30-year time frame.

Recommendation

The next generation of radars should be designed as part of an integrated observing system aimed at improving forecasts and warnings on relevant time and space scales.

THE CURRENT SYSTEM

The current NEXRAD system is a highly capable weather surveillance radar that has proved to be of great value to many sectors of our society, with its value extending beyond the traditional goal of protecting life and property. The Radar Operation Center (ROC) and the NEXRAD Product Improvement (NPI) Program are continually improving the system.

Early field testing of NEXRAD concepts and systems in a limited range of geographic and climatological situations did not elucidate and evaluate the full range of operational demands on the system. Weather surveillance needs vary from region to region and from season to season, and they depend on factors such as the depth of precipitating cloud systems and local topography. As the NEXRADs were deployed in other regions, further needs developed and additional limitations surfaced. The desire for more rapid update cycles is widespread, as are concerns about data quality.

Recommendation—Near-term

The Radar Operation Center and the NEXRAD Product Improvement Program mechanisms should be extended to permit continual improvement to the NEXRAD system. Provisions should be made to carry features found to be beneficial, such as polarization diversity, over to the succeeding generation of systems.

Recommendation

Weather surveillance needs should be evaluated by geographic region to determine if a common radar system design is appropriate for all regions.

Recommendation

The development program for the next generation weather surveillance radar system should incorporate adequate provision for beta testing in the field in locations with diverse climatological and geographic situations.

ADVANCED RADAR TECHNOLOGIES—
CAPABILITIES AND OPPORTUNITIES

The emergence of new radar technology provides an important foundation for updating the current NEXRAD system. A key technological issue related to future radar development and usage is that of spectrum allocation. Communications and other users of the electromagnetic spectrum are competing for the current weather radar spectrum allocation. There is particular concern that the use of S-band (10–cm wavelength) may be lost for weather radar applications. The loss of S-band would compromise the measurement of heavy rain and hail, the ability to provide warning of flash floods and tornadoes, and the monitoring of hurricanes near landfall. The cost of rectifying these impacts in the current NEXRAD system would be high, and the constraints on the design of a future replacement system would be serious.

Emerging hardware and software technologies offer promise for the design and deployment of a future radar system to provide the highest-quality data and most useful weather information. Adaptive waveform selection and volume scan patterns are important for optimizing radar performance in different weather situations. Radar systems with phased-array antennas and advanced waveforms can support a broad range of applications with observation times sufficiently short to deal with rapidly evolving weather events such as tornadoes or downburst winds. Polarimetric techniques offer means of dealing with many data-quality issues, provide a means for identifying hydrometeors over a storm, and offer the potential for the more accurate estimation of rainfall and the detection of hail. The ability of phased-array antennas to provide the requisite polarization purity has yet to be established.

Recommendation—Far-term

Adaptive waveform selection, which may even be applied to present systems, and agile beam scanning strategies, which require an electronically scanned phased array system, should be explored to optimize performance in diverse weather.

Recommendation—Far-term

The technical characteristics, design, and costs of phased array radar systems that would provide the needed rapid scanning, while preserving important capabilities such as polarization diversity, should be established.

Recommendation

The quality of real-time data should receive prominent consideration in the design and development of a next generation weather surveillance radar system. Real-

time data-quality assessment should be automated and used in deriving error statistics and in alerting users to system performance degradation.

Recommendation

Policy makers and members of the operational community should actively participate in the arena of frequency allocation negotiation. The impact, including the economic and societal costs, of restrictions on operating frequency, bandwidth, and power should be assessed for current and future weather radar systems.

NETWORKS AND MOBILE PLATFORMS

The new technologies also provide a foundation for the networking and placement of future radar systems. A closely spaced network of short-range radar systems would provide near-surface coverage over a much wider area than the current NEXRAD system. This would expand geographic coverage of low-level winds, precipitation near the surface, and weather phenomena in mountainous regions. Radars other than those in the primary network (e.g., weather radars operated by television stations or air traffic control radars operated by the FAA) could fulfill some of these roles.

Mobile radars can provide highly detailed views of weather events. Such observations not only have scientific interest, but also could be valuable in support of emergency services in cases such as fires, contaminant releases, and nuclear, chemical, or biological attacks upon this country.

Satellites and other aerospace vehicles represent alternatives to traditional ground-based systems. For example, the satellite-borne Tropical Rainfall Measurement Mission (TRMM) radar has demonstrated the ability to observe precipitation over regions not reached by land-based radars. Future satellite technology is likely to allow on-orbit operation of radar systems with larger antenna apertures and higher power outputs than are currently used in space. Satellite constellations, operating as distributed array antennas, could provide high-resolution global coverage. Both piloted and unmanned aerospace vehicles (UAV) are being developed for a variety of remote sensing and other applications. As the capabilities of these airborne platforms increase, it may become possible to place weather radar systems on station at a variety of altitudes, for an extended duration.

Recommendation—Near-term

The potential value and technology to incorporate data from complementary radar systems to provide a more comprehensive description of the atmosphere should be investigated

Recommendation—Near-term

The potential of operational mobile radar systems to contribute to the nation's weather surveillance system for emergency response and for improved short-term forecasts should be evaluated.

Recommendation—Far-term

The potential for a network of short-range radar systems to provide enhanced near-surface coverage and supplement (or perhaps replace) a NEXRAD-like network of primary radar installations should be evaluated thoroughly.

Recommendation—Visionary

The capabilities of future space-based radar systems to supplement ground-based systems should be determined.

Recommendation—Visionary

The capabilities of Unmanned Aerospace Vehicles and piloted aircraft to carry weather radar payloads should be monitored for their potential to provide weather surveillance over the continental United States and over the oceans.

AUTOMATED AND INTEGRATED PRODUCTS

Weather radar data are being increasingly used not only in forecasting and warning applications but also in climatological studies as well as in a wide variety of other research areas. Weather radar provides observations on the small space and time scales that are essential for monitoring precipitation and diagnosing certain weather events as well as for supporting nowcasting systems, hydrologic models, and numerical weather prediction models. Issues of data quality are central to most such applications, particularly to efforts to automate the applications. Effective assimilation of radar data in the models also requires detailed error statistics.

Broad dissemination of weather radar data in real time facilitates the application of these data to diagnostic and forecasting operations. Archiving of radar base data, as well as product data, facilitates research activities, retrospective studies, and climatological investigations. A long-term objective of the radar and other weather observation systems is the establishment of an integrated observational system, whereby most or all of these observations (e.g., ground-based, airborne, and space-borne radar, along with satellite, surface, and other data) would be assimilated onto a four-dimensional grid to provide the most complete diagnosis of weather impacts possible. Numerical weather prediction models and nowcasting techniques would then provide forecasts from a few minutes to many hours. A broad array of products will be used to support decisions that improve safety to humans, improve operational efficiency, and make homeland defense efforts more effective.

Recommendation

To support the use of radar data in the climate observing system and other research areas, standards for calibration and continuity of observations should be established and implemented.

Recommendation

The value of radar data, as part of an integrated observing system, in diagnostic applications, nowcasting systems, and hydrologic and numerical weather prediction models should be considered in the design of the next generation weather radar system. The characteristics of radar observations and associated error statistics must be quantified in ways that are compatible with user community needs.

Recommendation

Plans for next generation weather radar systems should include provisions for real-time dissemination of data to support forecast, nowcast, and warning operations and data assimilation for numerical weather prediction, and certain research applications. Routine reliable data archiving for all radars in the system for research, climatological studies, and retrospective system evaluation must be an integral part of the system. Convenient, affordable access to the data archives is essential.

Recommendation

Tactical Decision Aids and means for collaborative decision-making capabilities should be developed for both meteorological and nonmeteorological users of the system, with attention to the demands on the integrated observing system.

1

Role of Radar in the Weather and Climate Observing and Predicting System

Radars today are used to detect and track aircraft, spacecraft, and ships at sea as well as insects and birds in the atmosphere; measure the speed of automobiles; map the surface of the earth from space; and measure properties of the atmosphere and oceans. Principles of radar have led to the development of other similar technologies such as sonar, sodar and lidar (laser radar) that permit detection of phenomena and targets in the oceans and in the optically clear air.

In the past half century, weather radar has advanced greatly and has played increasingly important roles that span a wide spectrum of meteorological and climatological applications. Of particular importance has been its ability to detect and warn of hazards associated with severe local storms that include hail, tornadoes, high winds, and intense precipitation. Weather radar also monitors larger weather systems such as hurricanes that often include similar phenomena but can extend over very large areas. Today, weather radars improve aviation safety and increase the operational efficiency of the entire air transport industry, and they contribute to agriculture alerts and flood warnings through monitoring of rainfall intensity. They are also used regularly for recreational planning and other weather-impacted activities. Radar measurements have also been key to many remarkable advances in our understanding of the atmosphere and to better weather prediction over a variety of temporal and spatial scales. Such advances have been enabled through a combination of progressive improvements in radar hardware, signal processing, automated weather-based algorithms, and displays.

In recent years, added improvements in short-range forecasting and nowcasting have also resulted from the development of integrated observing systems that blend data from weather radar and other instruments to produce a more complete picture of atmospheric conditions. Two examples of such relatively new systems are the Advanced Weather Interactive Processing System (AWIPS)[1] and the Integrated Terminal Weather System (ITWS). AWIPS is a modern data acquisition and distribution system that gives meteorologists singular workstation access to NEXRAD radar products, satellite imagery, gridded weather forecast data, point measurements, and computer- and man-made forecast and warning products. The result is an integrated forecasting process that utilizes a comprehensive set of data for application by National Weather Service (NWS) Offices and others to generate more accurate and timely weather forecasts and warnings (Facundo, 2000). ITWS combines data from a number of weather radars, including NEXRAD, the Terminal Doppler Weather Radar (TDWR), and airport surveillance radars (ASR), with lightning cloud-to-ground flash data and automated weather station measurements to produce a suite of products that display current weather as well as nowcast weather out to around one hour for use by air traffic controllers in the management of airport terminal operations (Evans and Ducot, 1994).

The evolution of weather radar in the United States has been marked by the development and implementation of a series of operational systems, including the CPS-9, the WSR-57, and the WSR-88D (NEXRAD). Each of these systems was a response to the recognition of new needs and opportunities and/or deficiencies in the prior generation radar. The CPS-9 (X-band or 3-cm wavelength) was the first radar specifically designed for meteorological use and was brought into service by the U.S. Air Force USAF Air Weather Service in 1954. The WSR-57 was the radar chosen for the first operational weather radar system of the NWS. It operated at S-band or 10-cm wavelength, chosen to minimize the undesirable effects of signal attenuation by rainfall experienced on the CPS-9 3-cm wavelength radar. The development of the WSR-88D was in response to demand for better weather information and resulted from advances in Doppler signal processing and display techniques, which led to major improvements in capabilities of measuring winds, detecting tornadoes, tracking hurricanes, and estimating rainfall. These remarkable new measurement capabilities were a direct consequence of many engineering and technological advances, primarily advances in integrated circuits, digital signal processing theory, and display systems, and these advances led to advanced research weather radars. Radar meteorology research has also played a critical role in these developments through the generation of new knowledge of the atmosphere, especially regarding cloud and precipitation physics, severe storm evolution, kinematics of hurricanes, and detection of clear air phenomena such as gust fronts and clear air turbulence. Such knowledge has greatly benefited the operational utility of weather radar, particularly through innovations, understanding, and testing of algorithms that process radar data into meaningful physical descriptions of atmospheric phenomena and weather con-

---

[1]A complete list of acronyms and their definitions is provided in Appendix B.

ditions. It was the combination of technological advances with new scientific knowledge that enabled the deployment of the NEXRAD system and ensured its success as a highly valuable weather observing system.

This history of the national weather radar system and the multiplicity of factors that influenced the development of NEXRAD into its present form is necessarily brief. Most importantly, it does not do justice to the many persons who contributed to the current state of the nation's NEXRAD system or to the numerous scientific and technological advances that have made the system (current and future) possible. It is not possible to adequately credit all those whose knowledge and skills have led to the current system. However, a number of recent review articles by Rogers and Smith (1996), Serafin (1996), and Whiton et al. (1998) provide a starting point for this analysis. Additionally, a number of books and monographs, including works by Battan (1959, 1973), Doviak and Zrnic (1993), Atlas (1990), Sauvageot (1992), and Bringi and Chandrasekar (2001), provide valuable insight. The American Meteorological Society (AMS) preprints of the Conferences on Radar Meteorology also provide a rich resource on related matters.

As was the case with prior generation radar, the WSR-88D has achieved many more goals than was anticipated at the time of its design. The WSR-88D was motivated largely by the needs for early severe storm detection and warning. In this regard it has proved to be remarkably successful (Serafin and Wilson, 2000) and has become the cornerstone of the modernized weather service in the United States (NRC, 1999). But many other important applications have emerged from experience with NEXRAD and through advances in the research community. Thus, needs and opportunities have expanded and limitations have been found (see Chapter 2). Among the primary new developments in recent years is radar polarimetry. This development allows for data-quality enhancements and improved accuracy in the determination of rainfall. This is consistent with the emphasis on quantitative precipitation estimation (QPE) and quantitative precipitation forecasting (QPF), which have been identified as one of the top priority goals in meteorology by both the U.S. Weather Research Program (USWRP) (Fritsch et al., 1998; USWRP, 2001) and the World Meteorological Organization (WMO) (Keenen et al., 2002). Another advance has been the measurement of air motion in the optically clear air, which provides important wind information fundamental to a variety of applications. A more recent development based upon the long-term behavior of precipitation systems (e.g., Carbone et al., 2002) emphasizes the climatic applications of NEXRAD data.

Moreover, it is no longer appropriate to use the radar network as a stand-alone system. One cannot overestimate the importance of using the radars as part of an integrated observing system. On regional scales, the combination of the primary radar with subsidiary radars, with satellite data, with automated meteorological measurements from aircraft, and with a network of ground-based meteorological instruments reporting in real time has led to advances in vital nowcasting applications of severe weather. Such applications include improving the accuracy of severe local storm warnings (including forecasts of storm initiation, evolution, and decay), providing reliable guidance for construction activities, providing better information on current and future road conditions, furthering the needs of the aviation system for improving safety and operational efficiency (both civil and military), and helping individuals plan recreational activities.

Recommendation[2]

The next generation of radars should be designed as part of an integrated observing system aimed at improving forecasts and warnings on relevant time and space scales.

---

[2]Recommendations in this report appear in italics. Those in bold-face deal with technological approaches that are deemed worthy of consideration in the development of the future replacement for the NEXRAD system; they are categorized as "near-term," "far-term," or visionary." The committee uses the term "near term" for those technologies for which the capabilities exist currently and could be implemented even before the development of the replacement NEXRAD. "Far-term" technologies are those that could be available within the time period covered by this report (25-30 years), though they will require continued scientific and technological development before they could be implemented. "Visionary" technologies are those that may or may not be ready for operational use within the 25- to 30-year time frame. The other recommendations deal with the processes by which the future system is developed and deployed.

2

The Current System

As a baseline it is appropriate to begin with a review of the existing system. The primary weather surveillance radar system operated by U.S. agencies is the WSR-88D (NEXRAD) system, consisting of about 150 nearly identical radars. Most of these radars were deployed over the United States and some overseas locations in the 1990s. Data from this system support activities of the National Weather Service (NWS), Federal Aviation Administration (FAA), and the Department of Defense (DoD); they are distributed to a wide variety of other users as well. Although there are some differences among the radars operated by the three agencies, the design is essentially uniform for all radars. A common design for all the radars helped ensure high reliability and performance while reducing maintenance complexity and life-cycle costs. The primary functions of the system are to provide measurements for monitoring and forecasting severe storms, developing flash flood warnings, and contributing to other hydrologic applications. In addition to these, the radar system has evolved into a critical tool that supports numerous other meteorological applications. Data from other radar systems can also augment the NEXRAD data set. These systems include the FAA's Terminal Doppler Weather Radar (TDWR) and short- and long-range surveillance radars [Air Route Surveillance Radar systems (ARSR) and Airport Surveillance Radar (ASR)], atmospheric wind profilers (Rich, 1992), and radar systems operated by television stations and other private entities.

THE NEXRAD NETWORK

Appendix A summarizes the technical characteristics of the WSR-88D radar system. Crum and Alberty (1993) and Serafin and Wilson (2000) provide additional background on the system characteristics. Coverage over the eastern two-thirds of the country is essentially complete, though significant limitations exist in coverage near the surface. There are some gaps in western regions, and the combination of high-altitude sites and mountainous terrain presents difficult problems in several areas (Westrick et al., 1999).

Surveillance of the atmospheric volume surrounding a NEXRAD site is provided through one of several available volume coverage patterns (VCPs). The VCPs summarized in Table 2-1 are commonly used. The "clear air" patterns cover the lowest layers of the atmosphere in 10 minutes and provide such things as wind profiles and indications of sea breeze fronts or storm outflow boundaries that could trigger convective activity. The "precipitation" and "severe weather" patterns cover the full depth of storm activity in 5 to 6 minutes and provide more frequent updates on evolving storms.

Primary Data and Derived Products

The NEXRAD is a pulse-Doppler system that measures three primary characteristics of the radar echoes: equivalent radar reflectivity factor, commonly referred to as reflectivity and designated by Z_e; Doppler (radial) velocity, designated by v or vr; and the width of the Doppler spectrum, designated by σ_v. These base data variables, derived in the radar data acquisition (RDA) unit, express the zeroth, first, and second moments, respectively, of the Doppler spectrum of the echoes. A value for each quantity is available for every "resolution cell" of the radar, as defined by its antenna beamwidth and the sampling rate along the beam axis (though the latter is constrained presently to no less than half the pulse duration).

These displays, together with products (summarized in Radar Operations Center, 2002) and results of the algorithms discussed below, are developed from

TABLE 2-1. WSR-88D Volume Coverage Patterns[a]

Scan Strategy	Number of 360° Azimuthal Scans	Number of Unique Elevation Steps	Elevation Range of Azimuthal Scans	Time to Complete (Min)
Clear air (short pulse)	7	5	0.5° to 4.5°	10
Clear air (long pulse)	8	5	0.5° to 4.5°	10
Precipitation	11	9	0.5° to 19.5°	6
Severe weather	16	14	0.5° to 19.5°	5

[a]The azimuthal scan at the two lowest elevation angles (three for clear-air long pulse) is repeated to permit one scan in a low-PRF surveillance mode (to map the reflectivity field) and another in a high-PRF Doppler mode (to measure radial velocities). At higher elevations, these functions are done during the same azimuthal scan (Adapted from Crum et al., 1993)

the base data in a radar product generator (RPG) unit. In addition, a series of computer algorithms operate upon the base data (some also incorporate auxiliary information such as temperature profiles), examining the echo patterns and their continuity in space and time in order to identify significant weather features such as mesoscylones, tornado vortex signatures, or the presence of hail. Outputs of these algorithms are displayed as icons superimposed on the basic radar displays or in auxiliary tables. The number and the variety of potential algorithms continue to increase as scientific knowledge about the relationship between echo characteristics and storm properties improves and the available computational resources increase.

Data Display, Dissemination, and Archiving

A "principal user processor (PUP)" associated with each NEXRAD installation, and numerous additional "remote PUPs," provided the initial data display capability. The PUP was essentially a mainframe minicomputer, with a monitor, that operated programs to generate displays from a rather limited set of possibilities. As computer technology has advanced, open-systems architecture is being implemented to replace both the RPG and PUP units. Thus, the "open RPG" (ORPG) will generate and display the various products as well as relaying the relevant data on for display on other systems.

Although the NEXRAD system of radars is of major value as a stand-alone weather-observing network, additional value is obtained through the integration of NEXRAD data with other weather observations [e.g., other radar systems, wind profilers, satellites, the National Lightning Detection Network (NLDN), or surface measurements] and associated analyses. A number of systems have evolved to accomplish the integration of weather observations, many of which are intended to support commercial and general aviation. Among these are:

- AWIPS   Advanced Weather Interactive Processing System
- CIWS   Corridor Integrated Weather System
- ITWS   Integrated Terminal Weather System
- OPUP   U.S. Air Force Open Principal User Processor
- WATADS   Algorithm Testing and Display System
- WSDDM   Weather Support to Deicing Decision Making

These and similar systems are expected to mature dramatically and grow in use over the next two decades as the science of meteorology and the technology of information processing and dissemination continue to advance and merge with social needs for improved weather information and forecasting.

Dissemination of NEXRAD data within the NWS is now handled through the AWIPS system. Equivalent systems support the FAA and DoD users of NEXRAD data. Dissemination to outside users, formerly handled by vendors, is now accomplished through the Base Data Distribution System (BDDS) with distribution over the Internet.

Archiving of the NEXRAD base data (the "Level II" data) has been accomplished by magnetic tape recording at the sites, with the tapes being shipped to the National Climatic Data Center (NCDC) (Crum et al., 1993). Experience has shown that only a little more than half of the national data set reaches the archive in retrievable form. Thus, the Collaborative Radar Acquisition Field Test (CRAFT) (Droegemeier et al., 2002) is under way to test the capability to transmit NEXRAD data over the Internet to NCDC and increase the fraction of retrievable data. A separate archive of a set of the derived products (the "Level III" data) provides basic data for such things as research, training, and legal inquiries.

USERS AND USES OF THE DATA AND PRODUCTS

The NEXRAD principal user agencies are the Department of Commerce (DOC), DoD, and Department of Transportation (DOT). The primary mission organizations within these agencies are the NWS,[1] the Air Force Weather Agency (AFWA),[2] the Naval Meteorological and Oceanography Command (NMOC) and the FAA.[3] NWS is responsible for the detection of hazardous weather and for warning the public about these hazards in a timely, accurate, and effective way. The Service also provides essential weather information in support of the nation's river and flood prediction program as well as in support of civilian aviation, agriculture, forestry and marine operations. The national information database and infrastructure formed by NWS data and products, can be used by other governmental agencies, the private sector, the public, and the global community. The AFWA provides worldwide meteorological and airspace environmental services to the Air Force, Army, and certain other DoD organizations. NMOC supports the Navy, Marine Corps and certain other DoD organizations. The primary missions of these DoD agencies are to provide timely information on severe weather for the protection of DoD personnel and property; to provide weather-related information in support of decision-making processes at specific locations; and to support military aviation. The FAA's responsibility requires the FAA to

---

[1]The National Weather Service (NWS) provides weather, hydrologic, and climate forecasts and warnings for the United States, its territories, adjacent waters, and ocean areas, for the protection of life and property and the enhancement of the national economy.

[2]AFWA provides technical advice and assistance to all agencies supported by the Air Force weather support system and is responsible for the standardization and interoperability of the total Air Force weather support system. It also assesses the quality and technical goodness of weather support and fields standard weather systems for the Air Force, Army, and Special Forces.

[3]The FAA's mission is to provide a safe, secure, and efficient global aerospace system that contributes to national security and promotes U.S. aerospace safety.

gather information on the location, intensity, and development of hazardous weather conditions as well as to provide this information to pilots and air traffic controllers and managers. The current mission of the principal user agencies—to protect life and property—is expected to remain the same in the future. It is the quality, delivery, and use of this service that will change over time and will need to be addressed in considering the radar system of the future.

The group of users has expanded dramatically to include, among others, a very large atmospheric sciences and hydrometeorological research community in universities and research laboratories throughout the world; other federal, state, and local governmental organizations and private sector providers; and distributors and users of weather and climate information gleaned from meteorological radar measurements and other associated products. The latter include data that are either taken or derived directly from radar measurements as well as information derived through intelligent integration of radar data with other measurements and analyses of weather events. The NEXRAD system has been, and continues to be, immensely valuable in providing weather observations to this vast and diverse array of data users.

Many other federal agencies now use and rely on weather radar data to help meet their operational and other responsibilities. Among these are the Federal Emergency Management Administration (FEMA), Environmental Protection Agency (EPA), Nuclear Regulatory Commission (NRC), Department of Energy (DOE), U.S. Geological Survey (USGS), Department of Interior (DOI), Department of Agriculture (USDA), Forest Service, Bureau of Land Management (BLM), National Oceanic and Atmospheric Administration (NOAA),[4] National Park Service (NPS), Federal Highway Administration (FHWA), U.S. Coast Guard (USCG), and National Aeronautics and Space Administration (NASA).

Weather data have become valuable to the operations of many State and other governmental agencies. These typically include organizations with the following designations: agriculture, environmental protection, conservation and natural resources, fish and game commissions, transportation, emergency management, and water resources.

Numerous academic programs within the university community work with NEXRAD weather data in their research; these include departments or programs that represent studies in the fields of meteorology, atmospheric sciences, climatology, physics, chemistry, air pollution, hydrology, earth sciences, geography, geology, transportation, civil engineering, electrical engineering, geophysics, signal processing, computer science, computer engineering, natural resources, agriculture, forestry, economics, transportation, aviation, environmental science,

---

[4]The NWS is part of NOAA, but there are many other organizations within NOAA that independently utilize and rely on weather radar data and related products (e.g., National Climatic Data Center).

and engineering. Many government agencies and other public and private sector organizations are also involved in many investigations in fields that utilize weather radar data as an essential resource for their investigations.

As access and understanding of the use of NEXRAD weather data have grown, so has the list of users of this information within the private sector. Examples include broadcasters, commercial aviation, agriculture, trucking, recreation providers, economic forecasters, the insurance industry, and energy companies.

SHORTCOMINGS OF THE SYSTEM

A variety of limitations impede the ability of the NEXRAD system to meet the needs of all the varied users. Some, such as the divergence of the radar beam with increasing range, are inherent to any radar system. Others, such as the inability to acquire data in small elevation steps during shallow winter precipitation episodes, can be overcome by rather straightforward hardware or software modifications (the latter will be facilitated by the greater flexibility afforded by forthcoming open-systems architecture). The ongoing program of research and development should provide at least partial solutions to some of the other problems. But the opportunity to introduce newer technologies in a subsequent generation of weather surveillance radar systems offers promise of even further improvements.

Serafin and Wilson (2000) provide a good summary of the recognized deficiencies of the NEXRAD system. Those that affect the primary variables directly include contamination by ground clutter, both that in the normal radar environment and that arising during anomalous propagation conditions, and the occasional impact of bird echoes on the Doppler velocity data. The problem of range-velocity folding, common to all pulse-Doppler radars, has proved to be quite serious in the NEXRAD system. Study of several techniques now underway should yield means of mitigating the range-velocity folding problem, and versions of those techniques may well be applicable to future systems.

Spatial coverage limitations are imposed in the first instance by the curvature of the earth. This limitation constrains the available coverage to minimum altitudes, which increase with distance from the radar site. The problem is exacerbated by obstacles in the radar environment, which constitute a radar horizon extending above 0 deg elevation angle. The NEXRAD scans are restricted to some maximum elevation angle (currently 20 degrees), mainly to provide an acceptable scan update rate (see below); the result is a "cone of silence" data gap above each radar site. The cone of silence and limited low-level coverage inhibit the value of NEXRAD data to aviation interests. A further constraint on the NEXRAD system, limiting the minimum elevation angle to no lower than 0.5 deg, adds to this difficulty. This problem is of special concern for radars at high-altitude sites in mountainous areas; such radars are often unable to sense significant precipitation occurring entirely below their minimum usable elevation angle. Similar difficulties arise in areas subject to intense precipitation from shallow cloud systems, such as places in the lee of the Great Lakes affected by lake-effect snowstorms. Coverage over coastal and adjacent waters, a special concern for hurricane-prone regions, is limited. Such regional variations raise questions about the ability of a universal radar system configuration to meet the requirements for weather surveillance in all locations.

Recommendation

Weather surveillance needs should be evaluated by geographic region to determine if a common radar system design is appropriate for all regions.

Regions devoid of data pose a difficult problem for the NEXRAD algorithms. The primary causes of data voids are beam overshoot, cone of silence near the radar, beam blockage due to obstructions, gaps in vertical coverage, low signal strength, data masking due to data corruption, and planned and unplanned outages.

Overshoot refers to the void caused by the elevation of the lowest beam above the surface, which results from a combination of the elevation angle of the lowest tilt and the curvature of the earth. The consequential data voids affect every product. There is an additional complication. In order to achieve fairly rapid volume scans, the usual practice is to use a scan strategy, which has fairly coarse vertical beam spacing. The result is coverage gaps in the vertical. The trade-off here is between accepting longer times for the volume scan, accepting larger vertical gaps (fewer tilts), or enlarging the cone of silence by limiting the elevation of the highest tilt. Some products are more tolerant of vertical gaps than others. Early termination of volume scans by NEXRAD operators seeking more rapid updates of low-level base data occasionally introduces additional voids in the high-level data.

Data voids resulting from overshoot, beam blockage, vertical gaps, and the cone of silence are determined by the geometry of the radar installation and by the scan strategy. With the absence of reflectors, the signal strength can become so weak that the wind velocity cannot be resolved. The result is an enlarged velocity data void. Some compensation is possible by changing the waveform or the scan strategy, as in the present clear air mode VCP, if the radar was designed with the flexibility to trade increased volume scan time for greater sensitivity. Consideration should be given to providing the flexibility to adaptively adjust the sensitivity of the next generation radar.

The update cycle of the NEXRAD system, or indeed of any mechanically scanning system, in conditions of rapidly evolving convective weather is a serious limitation. Measurement of the primary variables along any given beam direction requires some minimum dwell time, which is governed by the required precision of the measurements. The dominant dwell-time constraint is imposed by the required precision in reflectivity (Smith, 1995), and in the case of NEXRAD, relaxing the 1 dB requirement could moderate this required precision. However, the necessity of covering the full volume of the atmosphere will always restrict the available update intervals.

A variety of other concerns about data quality exist. The false alarm rates for many of the current algorithms are higher than desirable. The limited spatial resolution at long ranges impedes the ability to identify small-scale weather features such as small tornadoes. Except for VAD winds, wind products have not been reliable enough for introduction into numerical weather prediction (NWP) models, largely because of artifacts in the data stream. The reliability of operation in remote and unattended locations can be a significant concern.

A final deficiency concerns the NEXRAD precipitation estimates (e.g., Smith et al., 1996; Anagnostou et al., 1998), which are important to a variety of applications. The spatial and temporal resolution of the data are not sufficient for many flash flood situations. There is a fundamental problem in converting the measured reflectivities to precipitation rates, in that no universal relationship exists, and means for establishing the relationship appropriate to a given situation are not at hand. Higher reflectivities in the "bright band" region contaminate many precipitation products. Moreover, the coverage limitations discussed above mean that reflectivity data over most of the surveillance area are only available for altitudes some distance above the ground. Methods for projecting the reflectivity data down to the surface are the subject of much research (e.g. Seo et al., 2000) but have yet to be applied in NEXRAD. Modifications to NEXRAD such as a polarimetric capability should help with the precipitation-rate problem, but the vertical-profile problem will remain ubiquitous.

The national coverage, improved accuracy, and rainfall estimation capabilities of the NEXRAD system have advanced the practice of hydrologic forecasting and water resource management. If dual-polarization capabilities are incorporated into the current system as planned, further improvements in precipitation measurements will occur, particularly in the quantification of high-intensity rainfall rates and the characterization of snowfall. A key to the success of any future radar system will be the preservation of capabilities such as dual polarization to provide improved data quality and high-accuracy rainfall measurements. Improved spatial sampling will be needed for representative near-surface coverage throughout the continental United States (CONUS) (possibly excluding regions in complex, highly mountainous terrain). The latter implies the need for an affordable means of dealing with the inherent inability of any widely spaced network to provide adequate near-surface surveillance over large portions of the country.

THE EVOLVING NEXRAD SYSTEM

Most of the field-testing of NEXRAD concepts and prototype systems prior to deployment took place in the central U.S. and dealt with warm-season convective weather. As the NEXRADs were deployed in other regions, further needs developed, and limitations that had not been elucidated in the earlier field-testing surfaced. The established mechanism of an Operational Support Facility (now the Radar Operations Center), advised by a Technical Advisory Committee, was able to deal with many of these concerns. Through this mechanism, algorithms have been revised, new ones have been added, and such capabilities as Level II data archiving have been implemented. But in hindsight, more comprehensive field testing covering a wider variety of regional and climatic conditions earlier in the system development process would have revealed some of the concerns soon enough to allow earlier and more effective action to mitigate their impacts.

Recommendation

The development program for the next generation weather surveillance radar system should incorporate adequate provision for beta testing in the field in locations with diverse climatological and geographic situations.

The WSR-88D system configuration is not static, but rather continues to evolve through an ongoing NEXRAD Product Improvement (NPI) program. Stated objectives of this program (Saffle et al., 2001) are to:

- ensure the capability to implement advances in science and technology to improve forecasts, watches, and warnings,
- minimize system maintenance costs, and
- support relatively easy upgrades in technology so that a large-scale WSR-88D replacement program may be indefinitely postponed.

The NPI Program works to develop and introduce system improvements in an orderly and seamless manner. To the extent that this program succeeds, the need for a replacement radar system will recede further into the future. Moreover, the NEXRAD system a decade or two hence will be substantially improved over that of today.

The current NPI Program emphasizes two major thrusts. One is to replace the data acquisition and processing systems in the original WSR-88D with "open-system" hardware and software. This development will increase the overall capability of the NEXRAD system for data processing, display, dissemination, and archiving; will facilitate implementation of new algorithms for processing the radar data; and will reduce costs for system operation and maintenance. Field deployment of the ORPG component, which executes the NEXRAD algorithms and produces the image products, should be completed in FY2002. Introduction of the second component, the open RDA (ORDA) unit, is scheduled to follow in about three years. The ORPG improvements provide a capability for (1) data quality improvements, such as AP clutter detection/suppression and identification of nonprecipitation echoes, (2) new polarimetric-based products such as improved precipitation estimation and hydrometeor particle identification, and (3) new products that may be directly assimilated into operational numerical models. The ORDA improvements will include (1) availability of a modern Doppler spectral processing platform including digital receivers for improved data fidelity, (2) a provision for range-velocity ambiguity mitigation techniques using phase coding and dual PRT waveforms, (3) capability for polarimetric sensing and processing to more accurately measure hydrometeor properties, such as drop size distributions and precipitation phase, and (4) custom VCPs to allow site-specific volume scans adapted to local weather needs. These improvements are expected to be operational in the next 5–10 years. The third major component of the WSR-88D, the PUP display unit, is being converted to open architecture in different ways by the three major NEXRAD user agencies.

The second major thrust of the NPI Program is directed toward introduction of a polarimetric capability for the WSR-88D. Such a capability could provide improved precipitation measurements as well as new capabilities for identifying hydrometeor types (e.g., recognizing the presence of hail or discriminating between rain and snow regions) and enhanced ability to screen out artifact echoes such as those caused by ground clutter or birds. The Joint Polarization Experiments (JPOLE) project planned for 2003 (Schuur et al., 2001) will evaluate these capabilities on a WSR-88D specially modified to provide a prototype polarimetric capability. Results of this project will influence the decision concerning full implementation of a polarimetric capability for the WSR-88D.

The NPI Program is not limited to modifications of the WSR-88D itself. An enhanced software environment, termed "Common Operations and Development Environment" (CODE), is being provided to facilitate use of the open-architecture system capabilities and linkage of the NEXRAD data to other agency weather data systems such as AWIPS. Plans and procedures are being developed to incorporate data from appropriate FAA radar systems, including the TDWR, ASR-9/11, and ARSR-4, as a means of expanding the available coverage and enhancing the backup capabilities in case of a WSR-88D outage.

The NPI Program provides a means for introducing continuing improvements in science and technology into the NEXRAD system on an ongoing basis. This evolutionary approach works to keep the system up to date, and can postpone the need for a replacement system until obsolescence issues such as mechanical wear and tear enter the picture. That can allow time for more complete evaluation of the various new technologies discussed elsewhere in this report. The NPI Program will also provide operating experience with new technical features, such as range-velocity ambiguity mitigation techniques or polarimetry, to help determine those features that should be carried over into a successor system. Sustaining the NPI Program will thus offer benefits not only to the NEXRAD system, but also to the follow-on system. Indeed, plans for the future system could well include a similar program for continual evolution.

Recommendation—Near Term

The Radar Operations Center and the NEXRAD Product Improvement Program mechanisms should be extended to permit continual improvement to the NEXRAD system. Provisions should be made to carry features found to be beneficial, such as polarization diversity, over to the succeeding generation of systems.

3

Advanced Radar Technologies: Capabilities and Opportunities

This chapter examines technological issues that are central to the concept of a longer-term technological view of a national weather radar system. The results are intended to extend 20–25 years into the future. The approach to the technology assessment centers on a new Radar Data Acquisition (RDA) system consisting of four key elements: the transmitter, the receiver, the antenna, and the processor—i.e., the hardware and software that produce the "base data" from which all relevant weather products are derived (also termed the "Level 2" data stream in NEXRAD). The discussion focuses on the most promising technologies for the NEXRAD replacement weather radar system.

The presentation is organized in four major sections. The first deals with a fundamental requirement of the NEXRAD system—preservation of the system's integrity through retention of the enabling frequency allocation resource for weather radars. The remaining sections deal with valuable improvements over the current system and describe several related technologies. Topics include means of improving data quality to reduce interpretive uncertainties, quantifying precipitation and improving precipitation classification via polarimetric measurements, use of phased array antennas to reduce volume sampling times and enable agile beam steering, and use of innovative signal processing schemes to improve radar performance.

Two fundamental concerns must be resolved before new advanced radar designs using promising modern technologies may proceed. First, critical technical issues need to be addressed related to the presently high cost of transmit-receive elements inherent in a phased array radar compared to the potential benefit in scanning flexibility and system reliability. It is not clear whether solid-state transmitter amplifiers (either within the individual modules or in a single transmitter configuration) offer enough advantages over high-voltage tube amplifiers to justify their inherent waveform constraints. Second, a political and economic debate continues between the communications industries (wireless and satellite radio) and the federal government regarding frequency allocation issues. It is clear today that future generations of mobile communication will apply pressure for expanded use of the existing S-band (10-cm) spectrum now used by ground-based weather and aviation radars. These potential revised spectrum allocations could limit or preclude radar operation in any of our existing "weather bands," in particular the low-attenuation S-band.

FREQUENCY ALLOCATION

Retaining the allocation of weather radar frequencies, particularly at S-band wavelengths with their relatively small attenuation in heavy rainfall, is critically important to maintaining the operational integrity of the NEXRAD radar system as well as preserving its capabilities for any future replacement system. Furthermore, an important advancement in the replacement system should be realization of more rapid volumetric sampling rate. This will most likely require use of radar waveforms that employ wider bandwidth than the current NEXRAD system. These frequency allocations are continuously being scrutinized for possible reallocation to other sectors of society to meet increasing spectrum demands for communications and entertainment applications. The issue is international in scope, since many spectral-use applications extend beyond national borders. Regarding the latter, the International Telecommunications Union sponsors the World Radio Committee (WRC) meeting in Geneva every two years to discuss and decide on spectrum allocation issues. Thus, in addition to preserving these national allocations, U.S. representatives to the WRC meeting must constantly remain alert in order to preserve the global weather radar spectrum allocations (McGinnis, 2001). The importance and value of weather radar to society has been aptly demonstrated with the NEXRAD system, especially through its utility for providing critical warning of severe weather, monitoring precipitation, and helping ensure safe and efficient operation of the National Airspace System (NAS).

The present weather radar users—e.g., the NWS, the FAA and the DoD—must make convincing arguments to spectrum allocation authorities to preserve the S-band region for weather activities necessary to aviation safety, preservation of life and property, and even national security, especially in light of the current world terrorist threat. These arguments can most readily be based on the ability of the longer-wavelength S-band radar systems to penetrate heavy precipitation and allow the proper interpretation of hydrometeor scattering without the complications that arise when the hydrometeor sizes are large relative to the radar wavelength.

Recommendation

Policy makers and members of the operational community should actively participate in the arena of frequency allocation negotiation. The impact, including the economic and societal costs, of restrictions on operating frequency, bandwidth, and power should be assessed for current and future weather radar systems.

DATA-QUALITY ENHANCEMENTS— POLARIMETRY AND AGILE BEAMS

Collecting and processing base data (the radar reflectivity, the radial velocity, and the spectrum width parameters) and the derived diagnostic (meteorological) products provide the bulk of the operational experience with NEXRAD. These experiences reveal data-quality problems. These problems are being addressed in open-systems architecture activities. The problems should continue to be addressed in the future system. The impacts of data deficiencies on specific products are described at length in a previous NRC report addressing NEXRAD coverage (NRC, 1995) and in Serafin and Wilson (2000).

Data corruption usually results from such factors as range folding, normal and anomalous propagation ground clutter, velocity aliasing, radio frequency (RF) interference, improper maintenance procedures, and nonatmospheric reflectors such as birds or chaff. Depending on the situation, the impact of these artifacts on generating an accurate meteorological product varies between minimal and severe. Product degradation can take the form of an enlarged data void when contaminated data are detected and censored, or it can take the form of erroneous products when biased data are passed on to meteorological algorithms.

Experience has shown that the integration of data-quality analysis prior to data assimilation is an effective way for detecting and masking erroneous data, thereby preventing the introduction of faulty information into the product algorithms. An automated data-quality analysis system should be an integral component of the next generation radar system. The primary component should be automatic detection of known artifacts and flagging of that data for special treatment prior to generation of any products using the radar base data. Certainly, these data-quality issues must be addressed within the data assimilation scheme if not sooner. Even more important for proper data assimilation is the knowledge of error statistics of each data source. Not only must the instrumentation error be known, but also the representativeness error of the specific measurements must be estimated for effective assimilation by a numerical model.

Recommendation

The quality of real-time data should receive prominent consideration in the design and development of a next generation weather surveillance radar system. Real-

time data-quality assessment should be automated and used in deriving error statistics, and alerting users to system performance degradation.

Although several studies are underway to address these important data-quality issues, two technological developments can provide a major improvement—polarimetric observations and electronic, agile beam scanning.

Polarimetric techniques

Tests on polarimetric radar tests are already being performed as part of a potential WSR-88D upgrade in the next decade. Polarization diversity observations bring some unique characteristics that are important for addressing data-quality issues. First, without any additional effort, polarimetric measurements automatically suppress the second trip echoes by about 15–20 dB depending on the type of hydrometeors. Second, the precipitation back scatter at horizontal (H) and vertical (V) polarizations exhibits a high degree of coherency (>0.98 in rain) that can be used to detect and filter contamination from noise as well as from nonmeteorological echoes such as surface clutter, chaff, birds and insects. Third, the differential polarization parameters, such as differential reflectivity and specific differential propagation phase, are immune to absolute calibration errors. Furthermore, self-consistency constraints of the covariance matrix measurements in rain impose bounds on errors in absolute reflectivity measurement (Scarchilli et al., 1995).

Dual-polarized radar systems can be configured in different ways depending on the measurement goals and choice of orthogonal polarization states. Fully polarimetric radar measures the complete covariance matrix of precipitation in the resolution volume (Bringi and Chandrasekar, 2001). Radars can be operated with polarization agility where the transmit polarization is changed on a pulse-to-pulse basis and two orthogonal polarizations are received, providing polarization diversity on reception. Alternatively, radars can transmit and receive the same polarization states, or utilize a hybrid mode in which they are different. The hybrid mode of operation where both horizontal and vertical polarization states are simultaneously transmitted but separately received is the mode being considered for the current WSR-88D upgrade. One drawback of the hybrid mode is that it inhibits the cross-polarization measurements that are extremely useful for water/ice discrimination and bright-band detection, though similar information can be obtained through the other polarimetric measurements.

Precipitation Classification and Quantification

Current NEXRAD-based products for the estimation of precipitation rate are seriously deficient. The deficiencies have been identified as a limiting factor for the value of the NEXRAD system in support of hydrologic products, including flash flood warnings. It is anticipated that the addition of a polarimetric capability to the NEXRAD will address, in part, these limitations. Dual-polarization measurements allow improved accuracy in the rainfall determination, more effective hail detection, and an effective means for characterizing hydrometeors throughout a storm volume. The improved accuracy in the determination of rainfall arises from the inclusion of differential reflectivity (Z_{dr}), which is an effective estimator of drop size, and specific differential phase shift (K_{dp}) as additional radar parameters to supplement reflectivity factor (Seliga and Bringi, 1976, 1978; Sachidananda and Zrnic, 1986; Bringi and Chandrasekar, 2001). The polarimetric radar will also permit the development of precipitation particle type products, an important addition to the diagnostic product suite (Vivekanandan et al., 1999; Liu and Chandrasekar, 2000). Plate 1 shows an example of vertical sections of reflectivity and hydrometeor type obtained from polarimetric radar. Independent of the impact on precipitation estimates, polarization diversity capability will contribute significantly to improving data quality (Chandrasekar et al., 2002). The next generation radar system should preserve the polarimetric capability anticipated for the WSR-88D, and special attention should be given to the development of products to quantify precipitation rate and to determine precipitation type.

The present inability of the WSR-88D to identify the bright band, and the inadvertent introduction of bright-band reflectivity bias into precipitation and storm products, has been identified as a recurring data-quality problem in existing NEXRAD products. The introduction of a polarimetric capability will allow for the development of high-quality bright-band information that can be used to compensate biased rainfall estimates. Polarimetric measurements offer a significant contribution to improved radar data quality, and the polarimetric capability is important to preserve within the framework of future radar systems.

Agile Beam Techniques

Other new capabilities of future radar will likely include much faster volume scans, or Volume Control Pattern (VCP) updates. Data quality can be improved using the precise steering of phased arrays to minimize clutter echoes. Rapid beam steering combined with flexible waveforms will offer new opportunities to minimize second trip echoes and to prevent velocity aliasing. All of these new capabilities will require much higher levels of digital processing power and storage than today's processors provide. However, these new digital processing and data storage requirements are not likely to limit the performance of the sensor systems described here. In addition, increased use of optical and wireless communications between the individual sensors and networks is envisioned.

Advanced radar technology can reduce, but not completely eliminate, data corruption due to ground clutter. Electronically scanned phased array radar has the flexibility to suppress ground clutter entering the main beam by steering the main beam immediately above clutter objects appearing on the horizon (mountains and buildings) and by skipping over known interference sources (such as the rising and setting sun). Further, electronic scanning allows complete removal of beam spreading and degradation caused by beam motion during the data acquisition interval, thereby yielding improved clutter rejection of both normal and anomalous propagation-induced clutter. The inherent beam agility allows rapid steering in other directions when objectionable interference beyond the primary echo is being received. Various data impairments associated with terrain blockage, beam divergence with range, overshooting weather caused by the earth's curvature, and atmosphere-induced beam propagation effects will continue to be present, but the overall data-quality improvement should be substantial.

PHASED ARRAY RADARS

Architectures

Both the FAA Terminal Area Surveillance System (TASS) study (Rogers et al., 1997) and the European COST-75 action (Collier, 2001) concluded that multiple-face phased array radar having no moving parts would not be economically feasible using the active antenna array module technologies currently envisioned. Several thousand transmit/receive (T/R) modules per face and at least four faces for full volumetric coverage requires on the order of 10,000 elements per radar. The total component cost may be greater than $10 million per radar, which seems an unlikely expenditure in the near future. The cost for a single-frequency dual-polarization T/R module is optimistically a few hundred dollars. Adding a second wavelength sensing capability would likely raise the cost to several hundred dollars per module. Benefit-cost studies will continue, but it appears that a major new T/R module design or array configuration is the only viable way to construct a network of fixed, multifaced phased array radars for weather services.

An alternative architecture utilizing a rotating *single* array appears entirely satisfactory for the next generation weather radar. A simple rotating slotted waveguide array has been proposed for a one-dimensional elevation steering concept of a single agile beam (Smith, 1974; Keeler and Frush, 1983; Josefsson, 1991; Holloway and Keeler, 1993). In its simplest incarnation, the rotating array covers the azimuth region by slow mechanical scanning while the elevation region is covered by an array tilted back about 20 degrees and using standard rapid 1D electronic scan steering techniques. Plate 2 shows a conceptual schematic of such a system. The single beam may be rapidly scanned electronically in elevation using frequency or phase steering that requires on the order of only 100 phase shifters instead of several thousand. In combination with a high-resolution pulse waveform, range averaging of independent high-resolution samples permits accurate base data estimates in a short dwell time. In this manner, a large volume can be scanned in time intervals on the order of a minute with spatial resolution

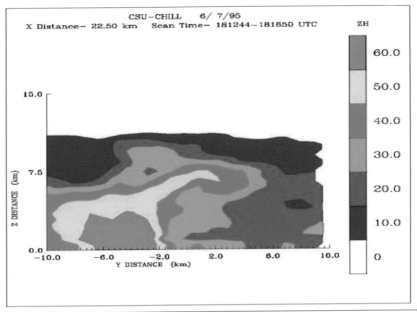

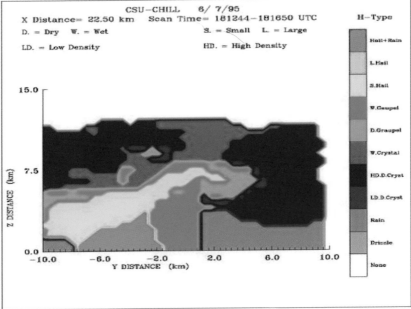

PLATE 1. Vertical section of reflectivity and the corresponding fuzzy logic based hydrometeor classification (adapted from Liu and Chandrasekar, 2000).

1

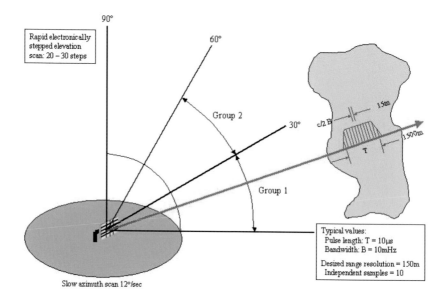

PLATE 2. Schematic diagram showing the concept of a single phased array antenna that scans mechanically in azimuth and electronically in elevation. In this example, two elevation groups allow coverage to 60 degrees using two rotations.

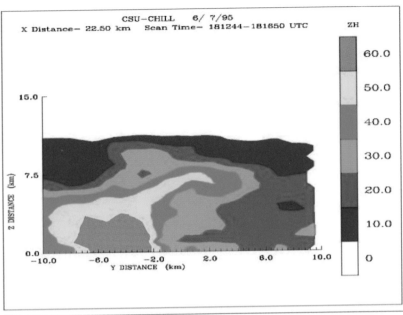

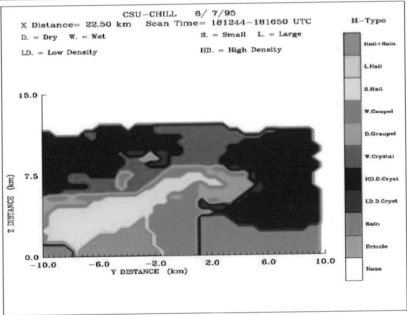

PLATE 1. Vertical section of reflectivity and the corresponding fuzzy logic based hydrometeor classification (adapted from Liu and Chandrasekar, 2000).

1

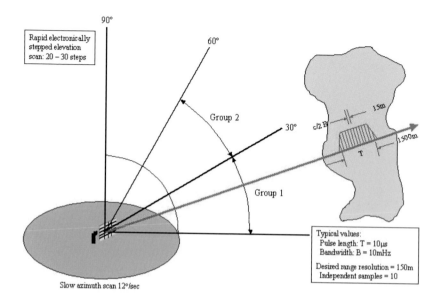

90°

Rapid electronically
stepped elevation
scan: 20 – 30 steps

60°

Group 2

30°

c/2 B

15m

1 500m

T

Group 1

Typical values:
Pulse length: T = 10μs
Bandwidth: B = 10mHz

Desired range resolution = 150m
Independent samples = 10

Slow azimuth scan 12°/sec

PLATE 2. Schematic diagram showing the concept of a single phased array antenna that scans mechanically in azimuth and electronically in elevation. In this example, two elevation groups allow coverage to 60 degrees using two rotations.

2

comparable to the WSR-88D. This approach requires increasing the transmit bandwidth, which will likely be complicated by future spectrum allocation issues. Typically, the phased array can electronically scan up to 20 degrees off a direction normal to the antenna plane without severely altering the beam shape. As the beam steers to higher elevations to cover the "cone of silence" above the radar, two degradations occur: (1) the beam size in that dimension is increased and (2) a larger number of elements are required (at closer spacing) to suppress grating lobes. It may be advantageous to execute a second mechanical sweep with the antenna tilted to a higher elevation and accept a corresponding loss in VCP update rate.

The FAA TASS program considered various other architectures, including back-to-back 1D scanning arrays, to increase the volume scan rate. A somewhat more complex and costly architecture utilizing 1D scanning in elevation but, in addition, having limited azimuth scan-back capability may be required for certain near-surface measurement applications requiring long dwell times, such as ground clutter cancellation or clear air detection of weak atmospheric signatures. Another architecture uses a small phased array feed, possibly only a linear array that illuminates a reflector that forms the main beam or beams. Digital beam forming (DBF) is feasible with some types of array feed systems, but beam shaping and sidelobe control may be difficult.

Multiple-beam phased array systems have been proposed (Skolnik, 2001; Hansen 1988). For example, multiple transmit beams, each having relatively high sidelobes using a high-power transmit antenna, could be coupled with a precision, low-power receive phased array capable of high overall isolation between the individual beams. Frequency steering techniques or digital beam forming technology would allow the same rapid VCP coverage while using longer dwell times on each beam. The primary disadvantage of multiple-beam architectures is the increased design effort necessary to assure high isolation between the simultaneous beams. Furthermore, the transmitter must generate a correspondingly higher average power (and a greater number of multifrequency waveforms in the frequency steered case) to maintain sensitivity. The same trade-off applies for single-beam techniques using pulse compression—the average transmit power must be increased to maintain sensitivity for a given range resolution.

Design tradeoffs in these alternative architectures are somewhat different. In the pulse compression single-beam system, the range sidelobes of the compressed waveform should be minimized to maintain the contrast in strong radial reflectivity gradients. In digital beam forming and frequency steered multiple-beam configurations using simultaneous transmit beams, advanced receiver beam shape design is needed to maintain isolation of nearby beams. Precision beam shaping may be achievable using low-power pin-diode phase shifters that are not required to accommodate the high-power transmit pulse.

Technologies

Phased array architectures may be based on different array technologies. The active antenna system uses individual T/R modules to integrate the transmitter, the receiver and the phase shifter with the antenna radiating elements. Examples include the Seimens-Plessey MESAR system and the Digital Terminal Area Surveillance System/Microburst Prediction Radar (DTASS/MPR) system (Protopapa et al., 1994, Katz et al., 1997). Active antenna technology has not been cost effective in the past, but future monolithic microwave integrated circuit (MMIC) designs and high volume may allow economical production for the next weather radar system.

Alternatively, a single transmitter may feed an array of radiators, such as phased array radars built by Lockheed-Martin Corporation (LMC)—the corporate feed SPY-1 radar (Maese et al., 2001) and LMC's newer space-fed Multiple Object Tracking Radar (MOTR) system. Recently, the University of Oklahoma and the National Severe Storms Laboratory in Norman, Oklahoma have made arrangements with the Office of Naval Research (ONR) and Lockheed-Martin to develop a mechanically scanned phased array using a Navy spare SPY-1 array face (Forsyth et al., 2002). This antenna will be coupled with a WSR-88D transmitter, a modern digital receiver and signal processing system, and (eventually) a pulse compression waveform. This phased array test bed, cumbersome by present standards, will be operated to test and evaluate phased array scanning techniques that will likely employ modern, cost-effective array technologies for a future implementation.

Slotted waveguide antenna technology offers a viable technology for phased array radar steerable in one dimension. Furthermore, the COST-75 action identified the Thomson-CSF reflect-array (Beguin and Plante, 1998) as a promising new technology for phase steering a single beam. When coupled with a high-range resolution waveform, these systems allow much faster volume update rates. Furthermore, antenna beam forming has analogies in spectrum analysis that may provide new insights for high-resolution antenna beams (Palmer et al., 1999).

The likely need to have polarization diversity integrated with the phased array places special constraints. TASS studies (Rogers et al., 1997) have shown that polarimetric transmission and reception is possible using the active antenna architecture whereby each radiator is capable of dual polarization. Studies at Ericsson (Josefsson, 1991) have shown that a slotted ridge waveguide can generate dual-polarization beams. However, these same studies indicate that the polarization becomes distorted when beams are steered away from the principal axes (only in azimuth or elevation from bore sight). Compensation may be possible since these are fixed, predictable patterns. The data-quality benefit will be attained even if the polarization purity deteriorates slightly in phased array antenna implementations when the beam departs from bore sight. These effects must be carefully considered in the design of any new radar system.

Because technology continues to evolve, it is important that there be further investigation and analysis of the risks, costs, and benefits that drive an advanced radar architecture decision. This analysis could lead to a conclusion that the costs and risks of the phased array technology are acceptable for the next generation system. In that event, research will be needed into the development of the appropriate prototype radar technology and processing algorithms for phased array systems. This research is necessary risk mitigation, so that appropriate information is available concerning options and benefits of a network of phased array radars and associated sensors.

Recommendation—Far-Term

The technical characteristics, design, and costs of phased array radar systems that would provide the needed rapid scanning, while preserving important capabilities such as polarization diversity, should be established.

Pulse Compression and Scan Rate

The volume update rate is a critical factor limiting the effectiveness of many meteorological products. Pulse compression utilizes a long, low peak power waveform coded in phase, frequency, or amplitude to effectively compress the full energy of the extended pulse to a much shorter resolution interval. Pulse compression techniques (or short, high peak power pulses) may be employed to acquire independent samples in range that may be averaged to obtain statistically accurate information in a short dwell time. For many aviation and military applications, an extended waveform using a form of pulse compression is entirely acceptable. However, for distributed weather scatterers, the integrated sidelobe contamination presents a large obstacle in the strong reflectivity gradients characteristic of convective weather. Various waveform designs have been proposed to reduce this contamination using nonlinear FM and other techniques of pulse shaping to reduce the Doppler sensitivity of the compression technique (Keeler and Hwang, 1995; Mudukutore et al., 1998; Bucci and Urkowitz, 1993). As other techniques of obtaining independent samples to reduce volume updates times are developed, these obstacles will be overcome. The wind profiler processing technique of range imaging (Palmer et al., 1999) provides independent samples in range similar to pulse compression. Inverse (whitening) filter processing techniques to increase the number of independent range samples (Keeler and Griffiths, 1978; Torres and Zrnic, 2001) may also find application to radar waveforms.

Solid-State Transmitter

In contrast to the distributed transmitters and receivers characteristic of active array technology, present microwave radars use klystrons, magnetrons, and traveling wave tubes to generate high-power pulses prone to single-point failures. Much has been written on solid-state transmitters regarding reliability and waveform flexibility. Even high average power solid-state transmitters are known to have high reliability and extremely low phase noise characteristics, primarily because they operate from low-voltage power supplies. They may be designed with failure-tolerant modes so that an entire second back-up transmitter is not required in operational systems as is the case for transmitters based on klystrons and traveling wave tubes (TWTs). However, solid-state transmitters lack the high peak power pulsing capability that may ultimately be required of a new generation weather radar system. Pulse compression of long, low peak power waveforms is a natural resolution; however, range sidelobe contamination in high-reflectivity gradients may limit the quality of the measurements. A definitive feasibility demonstration is needed for pulse compression techniques for weather radars before low peak power solid-state transmitters can be seriously considered for the next generation radars.

Dual-Wavelength Radar

A second shorter wavelength combined with an S-band system has been implemented within the research community for attenuation-based rainfall estimation and Mie scattering-based hail detection. The European community dropped this option after the COST-75 study cited potential practical problems, the major one being beam matching (Meischner, 1999). However, this technology is the one that is planned for the next space-borne radar system to be carried on the Global Precipitation Mission (GPM). In view of these conflicting pursuits, it is not clear whether dual-wavelength radar offers any additional benefit for future advanced weather radar.

Multiple Antenna Radar (Interferometric Processing)

Phased array radars offer another advantage over standard prime-focus reflector antenna systems. By separating the full array into two possibly overlapping subarrays in the receive mode and simultaneously processing the two beams, it is possible to retrieve the tangential wind component in the direction connecting the center of the two subarrays. This multiple antenna processing technique is well known in the vertical pointing wind profiler field (Briggs, 1980) and conceptually should be applicable to transverse wind measurements in an electronically scanned antenna operation using stationary beams. This processing application may require a slight azimuth scan-back to freeze the beam motion during the dwell time. By using digital beam forming techniques, all three components of the vector wind can be simultaneously measured at any radar measurement volume in space. In a vertical pointing mode, the profile of vector winds directly above the radar could be measured.

Global Positioning System (GPS)

The Global Positioning System (GPS) is a relatively new technology that continues to be exploited in atmospheric remote sensing. Several GPS-based retrievals are now available, and the suite of measurements will expand. In the context of radar networks, GPS has been used to provide a wide area coverage time and frequency synchronization between spatially separated components of various radar systems, such as the bistatic research radar systems (Wurman et al., 1993). The future network of radars will likely be synchronized by GPS. When networks of a variety of sensors must be synchronized over large distances and times, GPS offers an expeditious and inexpensive solution.

SIGNAL PROCESSING

Signal processing is an integral part of any weather radar. The technology and science of signal processing is fairly advanced and is not expected to be a limiting factor in the design of next generation radar. The technical advances and affordability in commercial signal processors, field programmable gate arrays (FPGA), and general-purpose computers are likely to far outpace the progress in other areas such as in transmitter and antenna design. In conjunction with signal processors, the receiver technology is also progressing at a rapid pace. Signal digitization is moving further up the receiver chain, and it is conceivable that future digitization may be performed after the first Low Noise Amplifier (LNA) and anti-aliasing filter. All these advances will permit advanced signal processing algorithms, such as adaptive filtering and spectral processing (Keeler and Passarelli, 1990). Some of these currently are being tested with research radars (Seminario et al., 2001). In summary, technological advances in signal processor and receiver technologies are expected to meet or exceed needs of the next generation weather radar.

Adaptive Waveform Selection and Scanning

Fully adaptive 2D beam steering capability allows extreme flexibility in developing waveforms not only to enhance data quality, but also to acquire rapid update data and to allocate transmit power where it is most needed. Because of the variety of applications and users, it is important to minimize the volume scan time and to use flexible scan strategies that adapt to varying weather situations. For example, much of the physical nature of convective storms resides in their vertical structure. Accordingly, vertical cross sections may be taken by adaptive scan and signal processing along arbitrary paths. Adaptive scan is also important in accelerating the scan cycle so that time is not wasted looking at empty space or low priority targets. Such a capability is critical in rapidly evolving situations such as tornado genesis and microburst formation, where a minute or two may mean the difference between life and death.

Agile beam electronically scanned phased array radars are capable of adaptive beam forming, adaptive scanning, and adaptive waveform selection to dwell on the most important regions of the atmosphere at any particular time. For example, a low-level surveillance scan might detect a region of convergence indicative of a potential thunderstorm near an airport. The radar could select a waveform for high clear air sensitivity, a beam shape that suppresses sidelobe clutter contamination from the ground and from more intense precipitation regions, and a scan pattern that monitors that region on a frequent revisit interval. The radar data system or products of the integrated observing complex can be used to determine the radar scans and waveforms on a constantly changing basis, depending on the evolution of the weather events in the region. Thus, the radar is not purely a data source; it actively feeds back information that it measures to enhance ongoing measurements. Furthermore, adaptive processing techniques that continuously optimize the sensitivity of the radar and suppress interference will be more frequently applied as these techniques are demonstrated in research systems.

Recommendation—Far-Term

Adaptive waveform selection, which may even be applied to present systems, and agile beam scanning strategies, which require an electronically scanned phased array system, should be explored to optimize performance in diverse weather.

Tomographic Processing

Tomographic processing has not been extensively applied to surface radars; however, it has extensive potential with path-integrated measurements (i.e., attenuation and differential propagation phase) to recover detailed atmospheric structures in the measurement plane (Dobaie and Chandrasekar, 1995; Srivastava and Tian, 1996; Testud and Amayenc, 1989). Tomographic processing may also yield path-integrated measurements from a single transmitter and multiple receivers, yielding a different paradigm for remote sensing of precipitation, and may find unique applications in airborne and space borne radar platforms. Nevertheless, tomographic processing is likely to remain in an auxiliary mode since it does not provide everything the current standard radar provides for all its applications.

Synthetic Aperture Processing

Synthetic aperture radar (SAR) systems have found widespread usage in space-borne applications, and the interest in using SAR systems for observing precipitation will grow continuously as additional airborne and space-borne missions are deployed. The SAR systems have the advantage of enhancing the spatial resolution with a smaller physical antenna (Fitch, 1988; Atlas and Moore, 1987). However, there are several special demands of weather radars that make SAR applications challenging. For example, most of the SAR operation is done at large incidence angles away from nadir, and ground clutter contamination is a serious problem under these conditions; motion of hydrometeors also forces a reduction in performance. Major innovation is needed prior to utilizing SAR technology extensively for space-borne observation of precipitation.

Refractivity Measurements

Conventional radar data processing is tacitly directed at precipitation, and the other echoes are "clutter" that must be suppressed. Yet, radar information is considerably richer. For example, anomalous propagation ground echoes reveal changes in the vertical profiles of temperature and moisture in the lower troposphere. Specialized processing is required in order to obtain and use these ancillary sources for meteorological information.

An example of such specialized processing is the measurement of the near-surface refractive index of air using fixed ground targets (Fabry et al., 1997). As the refractive index of air in the propagation path changes, the time of travel of radar waves between the antenna and the fixed ground target varies slightly and causes a change in the measured ground target echo's absolute phase. That variation is large enough to be accurately measured by carefully selecting and processing a large number of specific targets. If an independent set of temperature and pressure measurements are available, the refractivity can then be converted to spatial and temporal variations of moisture, or water vapor fields, in the surface boundary layer.

4

Networks and Mobile Platforms

The NEXRAD network radar-site selection was based upon a simple criterion—the aerial coverage at an altitude of 10,000 feet above the site level. A more comprehensive evaluation of radar coverage (NRC, 1995) measured the effective range of coverage as a function of specific weather phenomena. The 1995 NRC study illustrated convincingly that radar coverage by the NEXRAD network was generally far better than that of the network it had replaced. A few exceptions were found, and the agencies subsequently took remedial action.

By almost any measure, the network has proved its worth, serving the nation well and setting a new international standard for coordinated weather radar observations in a national context. Although the NEXRAD network was originally envisioned as an integral part of a complex and comprehensive integrated system in the modernized weather service, fully comprehensive integration of the radar information with other observational data in sophisticated interpretive algorithms and in data assimilation schemes with numerical models has yet to be realized. Much developmental work of this nature will certainly continue in the years to come, and weather forecasts will certainly be improved as a consequence of the NEXRAD data. There are, however, certain physical limitations that cannot be overcome with the existing deployment strategy. In the design of the successor network, these factors need to be considered seriously. However, it remains a fundamental assumption that the successor to NEXRAD will remain but one component in an integrated observing system. In addition to the primary radar network, complementary observations may include supplemental short-range radars, other government and private sector radars, mobile radars, space-based radars, and other observational systems.

AUXILIARY SHORT-RANGE RADARS

Limitations of individual, long-range radars are that at distant ranges, the beam overshoots the lower altitudes because of the earth's curvature, and the crossbeam resolution degrades progressively with increasing range.

A denser network of radars would provide solutions to many of these problems. Overshoot reduction and resolution enhancements are the obvious benefits. If the radars were close enough to provide coverage above their nearest neighbors, then the cone-of-silence voids would be eliminated. Because of the directional dependence of anomalous propagation, a dense network of radars might also provide some mitigation of these data corruption problems, provided that there is a data-level integration of the base data throughout the radar network. In addition to filling data voids, multiple viewing angles in overlap regions could provide multiple-Doppler resolution of the wind field. It is extremely important to investigate cost-effective ways to achieve a sufficiently dense radar network.

Thus, there are many reasons to consider a future radar system comprised of a network of long-range advanced radars operating at S-band and subnetworks of smaller radars operating at higher frequencies such as X-band. The latter would be used, for example, to cover the lower elevations below the radar horizon of the longer-range S-band radars, particularly in tornado-prone areas (where tornadoes have frequently been found to form at lower levels) (Trapp et al., 1999) and in mountainous regions. In addition, low-level boundaries such as coastal, sea-breeze, gust, and some synoptic-scale fronts may also be below the radar horizon. These boundaries are often the sites of significant weather events. Such short-range radars would also serve a variety of other functions. In particular, polarimetric X-band systems can possibly measure hydrologically and climatologically important light to moderate rainfall more accurately than is possible with S-band systems (Martner et al., 2001). These radars may also be useful in detecting conditions in clouds prior to the onset of lightning where electric fields align crystals vertically, leading to regions of negative differential phase shift that are readily detected by polarimetric radars (Caylor and Chandrasekar, 1996; Carey and Rutledge, 1998).

These X-based radars would be smaller and much less expensive than the larger S-band systems and would operate unattended with control from a central command office. Spread spectrum coding of the individual radar waveforms would allow them to operate in the same band with minimal interference. A relatively small fixed phased array, perhaps cylindrical or lampshade-shaped, could be used in order to allow coverage in all azimuths for a limited sector of elevations. Furthermore, line-of-sight attenuation measurements between the individual sensors over ranges of 20–50 km may provide additional information on precipitation rates or water vapor content using differential absorption techniques.

Anticipated advances in networking technology in the next decade will bring the cost and bandwidth of wide-area networking to levels that would make a high-density, highly coupled network of radars possible. Functionally, the radars would be sensors equipped with signal processing and display capabilities. The signals would be sent to a central location by fiber optic networks. The central facility would collect all the data in real time, process the data, and create the three-dimensional meteorological fields needed for forecasting and warning. Thus, with appropriate networking, a single radar could be divided into a system of smaller radars with short-range coverage that would enable new capabilities (Chandrasekar et al., 2001).

Figure 4-1 shows the concept of adding networks of small radars, similar to the concept of cell definition in wireless communication (Chandrasekar and Jayasumana, 2001). These smaller radars could be at a higher frequency and transmit less power than the S-band NEXRAD in use today. For the anticipated shorter range of coverage, the beamwidth could be wider while still maintaining a specific spatial resolution, since the coverage area would be reduced.

These multiple networked radars would be individually less expensive than the longer-range radar, but the total network cost might be comparable, even after including the added cost of developing the protocol and infrastructure for merging data from multiple radars. This concept can be extended to regions where terrain blockage is a problem. If a higher frequency were to be selected, then corresponding issues of attenuation correction would need to be addressed. Polarimetric techniques have been used successfully to correct the radar echoes for attenuation in rain (Bringi and Chandrasekar, 2001). However, the algorithms have yet to be demonstrated in regions of mixed-phase precipitation and have yet to be fully explored at X-band (and higher) frequencies. Because these smaller, higher-frequency radars would be expected to provide coverage over shorter ranges, the impact of path-integrated attenuation is not as significant as in longer-range systems. Therefore, the use of extensive networking to integrate large S-band phased array systems with smaller, short-range, short-wavelength radar, along with implementation of attenuation correction, could solve two critical problems of current-day systems. Specifically, the near-surface boundary layer coverage would be more uniform, and polarimetric capability would allow improved rainfall estimation at lower rain rates. These short-range radars could first be located in important rainfall and drainage basins, near airports, and near cities or other areas where boundary layer coverage is important.

Recommendation—Far-Term

The potential for a network of short-range radar systems to provide enhanced near-surface coverage and supplement (or perhaps replace) a NEXRAD-like network of primary radar installations should be evaluated thoroughly.

Some of the gap-filling function can be provided by more effective use of other operational weather radars. For example, the Terminal Doppler Weather Radar (TDWR) operated by the FAA and located at 45 major airports around the

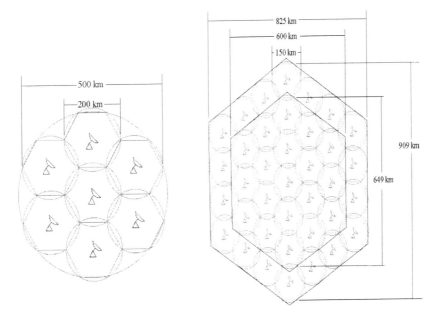

FIGURE 4-1. Comparison of widely spaced array of large radars (left) and closed spaced array of small, short-range radars (right) (Adapted from Chandrasekar and Jayasumana 2001)

nation provide superb low-level coverage in areas often not covered as well by the NEXRAD. There are also more than 100 weather radars operated by private sector organizations, primarily television stations that also could provide useful data.

Recommendation—Near-Term

The potential value and technology to assimilate data from complementary radar systems to provide a more comprehensive description of the atmosphere should be investigated.

RADAR PROFILERS

In the future, an integral part of the weather observing networks will be the radar profilers (operating at around 400 or 900 MHz) with radio acoustic sounding (RASS) for nearly continuous wind and temperature soundings. (The choice of radar frequency will depend upon frequency allocations.) The continuous measurement of winds and temperature is essential for a broad range of forecasting applications and for data assimilation into numerical models.

The lower-frequency profilers can also provide winds through the stratosphere. These have recently been found to have extended-range predictive value (Baldwin and Dunkerton, 2001).

Profilers have also been used extensively in vertical sounding of precipitation microphysics and kinematics (Atlas et al., 1999; Tokay et al., 1999). By providing a measure of the vertical air motions and the drop size distribution, the profilers allow one to improve microphysical models without the crude parameterizations that have been necessary in the past. Such information is also of considerable value in a variety of other contexts such as providing appropriate physically based Z-R relations. It can also be used as input to larger-scale models, and as means of calibrating conventional scanning radar. The profiling radars are a critical element of the future radar system.

OTHER COMPLEMENTARY OBSERVATIONS IN AN INTEGRATED NETWORK

Observing system networks currently contain a host of different approaches for making a variety of atmospheric measurements. Fundamental to these networks are in situ surface and radiosonde observations, the most basic of measurements used in numerical weather prediction (NWP) models and used for many short-term forecasts. Cloud-to-ground lightning flashes are now routinely monitored nationally, thus augmenting substantially severe weather and aviation forecasts. Polar orbiting satellite data are valuable in numerical weather prediction, and geostationary satellite data provide continuous images of clouds, water vapor, temperature, and other variables along with wind data. The increasing availability and sophistication of digital photographic and video equipment, coupled with the emphasis on increasing the data transmission bandwidth available to nearly everyone, means that a wide variety of visual imagery provided by amateurs as well as professionals may play a role in the future observing system. SuomiNet, a national network (Ware et al., 2000) is using ground-based and inexpensive global positioning system (GPS) receivers to measure water vapor fields. Further in the future, Constellation Observing System for Meteorology, Ionosphere, and Climate (COSMIC)— a constellation of low-orbiting satellites with GPS receivers (Rocken et al., 1997; Kuo et al., 2000)—will make thousands of soundings daily of atmospheric refractivity, which will be highly resolved in the vertical axis, and relatable to temperature and water vapor.

To date, however, the broad range of network data has not been exploited in a fully integrated fashion. In weather forecast offices, such supplemental observations are taken into consideration by individual forecasters, but usually without the benefit of advanced data assimilation analyses or high-resolution NWP guidance. Similarly, Doppler radar data are only now being evaluated for their potential to better initialize and improve the performance of high-resolution and national mesoscale models. In addition, a network of ground-based GPS receivers would provide horizontal fields of vertically integrated precipital water vapor. However, there is little doubt that research in the next decade will achieve considerable progress in the field of data assimilation and high-resolution NWP modeling for at least three reasons (Droegemeier et al., 2000). First, this is a topic receiving great emphasis in the U.S. Weather Research Program. Second, the topic is of considerable intellectual interest and is attracting outstanding people to the field. Third, digital computing technology is advancing so rapidly that the tools necessary for advanced operational modeling will be available and affordable (More detailed descriptions of these issues are found in Chapter 5 and in NRC 1999).

MOBILE RADAR

In recent years mobile and airborne radars have been used successfully in studies of severe storms and have been obtaining unprecedented high-resolution observations of tornadoes during their birth and evolution. Wurman and Gill (2000) made such observations with an X-band radar (Doppler on Wheels, or DOW); Bluestein and Pazmany (2000) provided equally impressive measurements with a mobile 3-mm-wavelength radar. In addition, Wakimoto and Liu (1998) used the X-band airborne Electra Doppler Radar (ELDORA) to document storm-scale motions in a super-cell while a tornado formed. All such observations have revealed significant aspects of tornadoes and their genesis not previously resolved. Richardson et al. (2001) have also obtained complete flow fields using dual-Doppler observations from two DOW systems. With the aid of local severe storm forecasts, groups from the University of Oklahoma have been able to position mobile Doppler radars near tornadoes as they developed and moved (Bluestein and Pazmany, 2000). This achievement suggests that it may be feasible for mobile radar units to be used to provide data to assist in storm warnings and to supplement warnings that are based solely upon the fixed-network radars. For example, real-time wind speed estimates in tornadoes and the precise location of the tornadoes could be relayed to the National Weather Service (NWS) and local communities.

Mobile Doppler radars could also be used to set up multiple-radar networks on the coastline or inland near to but a safe distance from land-falling hurricanes. In the past, the DOWs have been transported to coastal regions of North Carolina, the central Gulf, and Florida with adequate time to be set up in advance of hurricane landfall. Data from mobile radar networks could be used to determine more accurately the boundary layer wind field so that nowcasts and forecasts of surface wind speeds could be improved.

Ad hoc mobile Doppler radar networks could also be used near and downwind from sites of nuclear, biological, chemical, and other accidents or acts of terrorism. Multiple-Doppler wind fields could be used to determine areas most affected by contaminants, and they could also be used as input for very small-scale numerical models, which might be able to provide short-term forecasts of contaminant fields.

While mobile, short-wavelength radars could be used for special purposes in limited areas, transportable NEXRAD replacements could be used near NEXRAD sites at which the local network operational radars are not working for extended periods of time.

Recommendation—Near-Term

The potential of operational mobile radar systems to contribute to the nation's weather surveillance system for emergency response and for improved short-term forecasts should be evaluated.

AIRBORNE RADARS

The use of radar on aircraft as a weather detection and avoidance tool goes back approximately 50 years, when the airlines, particularly United and Northwest Airlines, worked with several manufacturers to develop the first airborne storm avoidance radars. In time, a requirement for on-board weather radar became a federal (and international) requirement. The military has developed several airborne radars that have weather detection capabilities, which have been used occasionally. Today, many of the commercial airborne radars have upgraded with Doppler capability, primarily to detect turbulence aloft.

Within the past several years, unmanned aerospace vehicles (UAVs) have become prominent, especially in the military, which has an operational need to "see" enemy activities in "denied" areas where more conventional observations of military operations are difficult. Another major use of UAVs has been the collection of weather data in remote, typically oceanic areas. It is envisioned that during the coming years, there will be two applications of airborne radar heretofore not addressed. During the coming years, unmanned aircraft operating at a variety of altitudes may be able to carry weather radars and view from above regions not accessible to the fixed ground-based network. For this to occur, UAVs, which proved to be critically important tools in Afghanistan, must be modified to include airborne radars; consequently, a strong case can be made for such a development. The current payload would be a difficult barrier to overcome. The solar-powered NASA HELIOS, developed by Dr. Paul MacCready (who first achieved human-powered flight with the Gossamer Condor), has been tested at altitudes above 80,000 feet and is expected to be able to gain 100,000 feet. The current system has such a limited payload (thought to be 150 pounds) that radar remote sensing would be prohibited, but the emphasis on UAVs may allow for an increase of payload, or unique ways may be found to develop very small radars suited to the platform. Some of these UAVs operate at low altitude, and small radar could be used. High-altitude UAVs like the jet-powered Northrop Grumman Global Hawk (a long endurance aircraft that flies above 60,000 feet) could be an excellent platform.

To fill vital gaps in precipitation observations over the oceans and to provide calibration of satellites over the oceans, consideration should also be given to the use of existing and new radars and radiometers on board intercontinental commercial aircraft. The data from existing weather radars on these aircraft would serve to fill in the data-void regions until precipitation satellites are able to cover the globe with adequate time and space resolution. Calibration of the airborne radars would also provide a transfer calibration to the space-borne systems.

There are many cases in which transcontinental aircraft fly above significant weather that goes unreported. At such times it would be useful to switch the radar to a vertically pointing antenna below the aircraft. In the near future it is expected that airborne radars will be coherent and capable of Doppler measurements. By beam switching the antenna fore and aft and left and right, it would be possible to obtain excellent measures of the wind profile as well as quantitative measurements of the precipitation.

It is also worthwhile to consider equipping commercial aircraft with synthetic aperture radar (SAR) and scatterometers. An airborne C-band scatterometer is in operation on the Gulfstream aircraft operated by the National Oceanic and Atmospheric Administration (NOAA) Office of Aircraft Operations (OAO). This latter system provides both measurements of the near sea-surface winds and velocity azimuth display (VAD) of the winds above the surface.

Recommendation—Visionary

The capabilities of Unmanned Aerospace Vehicles and piloted aircraft to carry weather radar payloads should be monitored for their potential to provide weather surveillance over the continental United States and over the oceans.

SPACE-BASED RADAR

The first satellite dedicated to the measurement of rainfall—the Tropical Rainfall Measurement Mission (TRMM)—was launched in November 1997 and is expected to maintain observations until 2007, thus establishing an important climatological record. TRMM has performed beyond all expectations and has contributed in many ways that were not anticipated prior to launch:

- TRMM has greatly enhanced estimates of the zonal distribution of rain over the tropics,

- Assimilation of data on rainfall and total precipitable water into TRMM has improved predictions of the hydrological cycle, clouds, and radiation in global computer analyses,
- TRMM has aided in the monitoring of the evolution of precipitation from the 1998 El Niño to the 1999 La Niña and the associated changes in storminess and precipitation around the tropics,
- TRMM has provided a single transfer standard for calibrating ground-based radars around the world.

Space-borne weather observations have the same advantages as satellite-borne platforms for all disciplines—coverage and communication. TRMM has proved the technology of space-borne weather radar in terms of stability, coverage, range/azimuth resolution, and observation window. However, the azimuthal swath is quite narrow (220 kilometers), so that it only overlaps about a third of the swath of the on-board microwave radiometer. This creates problems when one tries to "extend" the rain profile derived from the precipitation radar out to regions beyond the radar swath. Also, the long intervals between successive observations limit the utility of space-borne observations for weather prediction.

In addition to providing rainfall measurements, TRMM has also provided high quality measurements of the ocean surface cross section; the latter are also well related to the near surface winds. We therefore anticipate that a future radar-radiometric satellite will provide data on both precipitation and ocean surface winds.

It is expected that TRMM will be followed by the global precipitation mission (GPM), the orbit of which orbit will extend to extratropical latitudes ($\pm 60°$). It is planned that GPM will carry a dual-wavelength radar system for more accurate precipitation measurements. In addition GPM will fly a constellation of microwave radiometers to permit a three-hour revisit time, thus allowing the use of the data for weather prediction.

Significant advances in micro-satellite technology combined with radio frequency (RF), power, and receiver subsystems for space may provide an opportunity to bring space-borne surveillance radar within affordable levels in the next two decades. Indeed, Cloud Sat is another planned space-borne radar mission intended for launch in 2005.

Among the various technological advances will be the deployment of numerous micro-satellites that can fly in low earth orbits and can selectively provide continuous coverage for North America. This assumes implementation of other technologies that are under development for space applications: distributed-aperture systems, adaptive deployment, and spot beam or electric scanning. The COSMIC system, discussed earlier, is a nonradar example of such a constellation.

Multiple low earth orbiting satellite systems provide continuous coverage, whereas distributed aperture systems provide narrow footprints at subrun levels.

Spot beam and electronic scanning technology enable adaptive deployment. Although there are many advantages to satellite-borne systems, they still suffer a basic limitation of contamination by ground clutter at near-surface observations. In summary, space technologies may advance to a point where affordable space-borne weather surveillance radars can play an ancillary role in conjunction with ground-based radars.

Measurements of surface winds over the oceans by space-borne scatterometers and microwave radiometers have been made possible by satellite systems such as the NASA Scatterometer (NSCAT), by the scatterometers on the European Remote Sensing satellites (ERS-1 and 2), by QuickSat, and by the Special Sensor Microwave/Imager (SSM/I) on the Defense Meteorological Satellite Program (DMSP). The impact on global and regional forecast skill has been assessed by several organizations and is reviewed by Atlas et al. (2001). The key portions of their summary follow.

> Scatterometer data over the oceans are able to delineate precise locations and structures of significant meteorological features, including cyclones, anticyclones, fronts, and cols. As such, their use by marine forecasters can result in improved analyses, forecasts, and warnings for ships at sea and other marine interests. The use of scatterometer observations in data assimilation systems can extend their usefulness substantially and lead to improved sea level pressure analyses, improved upper air analyses of both wind and geopotential, and improved short and extended-range numerical weather forecasts. Typically, forecast improvements are skewed toward the Southern Hemisphere extratropics, with the length of a useful forecast extended by as much as 24 h in some Data Assimilation Systems. In the Northern Hemisphere extratropics forecast impact is close to neutral or slightly positive, but occasionally the depiction of an individual storm is improved substantially, leading to a more skillful forecast.

Finally, observations of the ocean surface by airborne and space-borne SAR have revealed heretofore unobservable atmospheric processes by virtue of the effects of the winds on the development of the short surface waves. The nature, intensity, and direction of storms have been deduced from the pattern of SAR echoes on the sea. SAR has also observed cyclonic circulations in the lower atmosphere prior to the development of hurricanes.

Recommendation—Visionary

The capabilities of future space-based radar systems to supplement ground-based systems should be determined.

5

Automated and Integrated Products

In the current NEXRAD system, automated products derived from radar base data provide important information directed toward specific needs and users. These products have provided valuable support to the forecast and warning process. Although the mechanisms discussed in Chapter 2 are providing continuing improvements in these products, operational experience has identified shortcomings. These shortcomings can be traced to the quality of the radar data and to the original NEXRAD requirements. Investigations into nowcasting and into the assimilation of radar data into numerical weather prediction (NWP) and hydrologic models have revealed additional demands on the radar data and the derived products. In this chapter, radar data issues will be presented from the viewpoint of the support needed for these automated analyses.

It is convenient to discuss the role of radar data in its support of different information horizons:

- diagnosis—a quantified description of the current atmospheric conditions,
- nowcast—short-term, precise forecasts of specific events (0–2 hours),
- short-term forecast—the prognosis of the state-of-the-atmosphere and specific events in the next several (2–6) hours, including products for the forecaster and for data assimilation,
- longer-range forecasts and support of climatological studies—the analysis of atmospheric information on longer time scales such as 6–48 hours and longer, and
- off-line product development.

Although the exact boundaries between the first three time horizons may blur, the distinctions are often meaningful from the viewpoint of current product development techniques. The ideal system would have seamless continuity, in which a fully implemented four-dimensional data assimilation (4DDA) system could render these distinctions obsolete. The integrated observing system should provide a comprehensive 4-D database from which various users can draw those portions that are pertinent to their applications.

RADAR COVERAGE AND DATA QUALITY

The following types of data limitations reduce the effectiveness of the current NEXRAD products:

- data voids,
- data corruption,
- degraded resolution at long range, and
- data latency (update rate).

Data voids have many causes (as discussed in Chapter 2), but the end result is the denial of desired data to a product algorithm in some portion of the coverage volume.

Data corruption usually results from a combination of factors, and the impact varies between minimal and severe. Product degradation can take the form of an enlarged data void, when contaminated data are detected and masked, or of erroneous product results, when incorrect data are passed on to the product algorithms. Advanced engineering techniques can reduce, but probably not totally eliminate, data corruption. Experience has shown that the integration of data quality analysis (DQA) with a data assimilation system is an effective way of detecting and masking erroneous data in order to prevent the introduction of faulty information into the product algorithms.

Data latency results from the use of a scan rate that is slow compared with the requirements of an effective product algorithm. Different products and different atmospheric situations impose different data update requirements. To alleviate data latency issues for all algorithms, it is necessary either to have a fixed scan strategy that is suitably rapid for all circumstances or to have an adaptive scan strategy that can concentrate radar resources effectively. Operational mechanical scanning radars are usually restricted to 360-degree scans, with each rotation taking 15–20 seconds. Faster scanning results in reduced sensitivity. Single-beam radar collects data from one tilt per revolution. Simple counting provides the relationship between the rotation time, the number of tilts, and the volume scan time. Continuing the current design would provide little room for improvement over the NEXRAD. Fundamental changes in the basic system, such as use of phased array technologies, would provide data updates at rates necessary for all critical products.

DIAGNOSTIC PRODUCTS

The original NEXRAD design focused primarily on requirements to address diagnostic severe storm products. Shortfalls in these products can be attributed primarily to coverage and data-quality issues. In addition, new applications have been developed that the original design did not anticipate. Overshoot and beam blockage are responsible for serious limitations in the observation of low-level precipitation and boundary layer winds. Surface winds, wind shears, and convergence lines are essential for the diagnosis and nowcasting of the evolution of convection and of the transport of pollutants and hazardous materials and for the diagnosis of damaging winds and aviation hazards. Special attention should be given to providing adequate coverage in the boundary layer.

NOWCAST PRODUCTS

The automated nowcast applications of the current NEXRAD are mostly tracking algorithms. In addition, human forecasters make extensive use of the radar products in support of their nowcast functions. Some prototype expert nowcast systems have been developed in research programs, especially in the areas of the growth of convective storms and quantified precipitation estimation (QPE). It has not been determined whether nowcast products will remain the responsibility of forecasters or whether they can be fully automated in the near future. Regardless of the nowcast agent, forecaster, or algorithm, there is a requirement for high-quality and comprehensive radar data.

The nowcast data requirements are the same as the diagnostic product requirements:

- Comprehensive and accurate boundary layer wind and reflectivity information is required for the successful nowcast of convective storm development and precipitation rate.
- Accurate precipitation-type identification is required for improved QPE nowcast.

FORECASTS AND ASSIMILATION OF RADAR DATA INTO NWP MODELS

There has been major progress in NWP of synoptic scales of weather, i.e., for spatial scales on the order of a few hundred kilometers or longer and for time scales of one day or longer. With the improvement of models and of methods of data assimilation, and with operational ensemble forecasting, such forecasts are now becoming routinely skillful for five days or longer. Because initial value problems are an important part of NWP, the improvement in the initial conditions for the forecast models is an essential component in this evolution.

There have been quite a number of experimental and operational studies on the assimilation of radar observations from NEXRAD in the United States and from other Doppler radars in Europe and Japan. Alberoni et al., (2001) provide a review of these efforts with an extensive reference list. We now review briefly the experience in assimilation of wind, water and precipitation data and indicate future requirements. Several methods for the physical initialization of models using estimated precipitation rates from satellite observations have been applied in global models in the tropics (e.g., Krishnamurti et al., 1993; Treadon, 1996; Falkovich et al., 2000). These methods change the model initial conditions during a period before the start of the forecast, forcing the model to produce rain where observed and eliminating it where satellites indicate no precipitation. Generally they result in improved short-range forecasts of precipitation, but the improvements do not last long. Regional experiments assimilating radar/rain-gage estimates of precipitation have also been successful in Japan (Matsumura et al., 1997) and have been implemented in the National Centers for Environmental Prediction (NCEP) Regional Reanalysis System (DiMego et al., 2001).

Rain and cloud water content can also be assimilated with relatively crude methods, using radar-precipitation relationships to estimate rain (e.g., Xue et al., 1998, Haase et al., 1999, Zhang, 1999). Results generally indicate an improvement due to a phase correction of the location of the convective systems Grecu and Krajewski (2000) used variational assimilation of radar data into models to develop rainfall forecasts. As indicated before, radar is the only observing system with the potential of providing initial conditions for very high-resolution numerical weather prediction models. The NWS has implemented a new 12-km-resolution Eta model at continental scale, and in the next decade a continued increase in resolution can be expected. The future radar should provide accurate information on winds and precipitation fields for these models, which should lead to significant improvements in the 6- to 72-hour forecasts. Information about the error statistics (e.g., Ciach and Krajewski, 1999; Keeler and Ellis, 2000) will also be needed for effective assimilation of the radar data into these models.

By comparison, it is particularly challenging to predict short-lived phenomena such as thunderstorms. One of the most promising and active scientific frontiers in numerical weather prediction is the prediction of severe weather (tornadoes, squall-lines, intense summer convection, etc.) with mesoscale or even storm-scale models. In order to predict the evolution of these phenomena, the models must have very high resolution (grid sizes on the order of 1–10 kilometers), comprehensive boundary layer and precipitation physics, and the ability to provide initial conditions with sufficient accuracy and comparable spatial resolution, especially in the boundary layer. The use of models initiated with high-resolution radar data offers promise for improving the skill of forecasting these short-lived phenomena. The three most important variables required for initialization of storm resolving models are:

1. horizontal winds,
2. water substance (including phase), and
3. temperature fields.

Radar provides high-resolution (space and time) information regarding the first two, and the third one can be indirectly derived from the winds.

A number of useful applications have already been demonstrated for use of powerful radars such as the WSR-88D in optically clear air. In such cases the echoes are due either to the windborne insects or to backscatter from turbulent fluctuations in refractive index (Wilson et al., 1994; Gossard, 1990; Serafin and Wilson, 2000). In either case one may use a single Doppler radar to obtain quite good measurements of the winds. Sun and Crook (1997a, b) have demonstrated a method for obtaining high-resolution wind and thermodynamic data from a single Doppler radar and a numerical model. Because a dense network of surface observing stations in Oklahoma have proved to be extremely valuable over a broad spectrum of human and industrial activities (Morris et al., 2001), such radar-derived wind and thermodynamic data have a wide variety of applications to the aviation, construction, agriculture, energy, and trucking communities and to hazardous chemical, biological, and nuclear accidents (Serafin and Wilson, 2000).

Among the techniques which have already shown promise for thunderstorm nowcasting are the Thunderstorm Auto-Nowcaster developed at the National Center for Atmospheric Research (NCAR) and the ITWS developed by MIT Lincoln Laboratory. These systems use radar, satellite, and conventional meteorological observations in expert systems for short-term forecasting. They have already proved to be valuable in air traffic control. When combined with storm-scale numerical models, they are also expected to be of great value in quantitative precipitation forecasts and flood warning (Serafin and Wilson, 2000). Subjective forecasting and nowcasting would also be greatly benefited by such datasets.

Recommendation

The value of radar data, as part of an integrated observing system, in diagnostic applications nowcasting systems, and hydrologic and numerical weather prediction models should be considered in the design of the next generation weather radar system. The characteristics of radar observations and associated error statistics must be quantified in ways that are compatible with user community needs.

LONGER-RANGE FORECASTS AND CLIMATOLOGY

Radar climatology involves classification of the historical patterns of atmospheric events as they were observed by the radar system. Issues include event frequencies, geographic distributions, distributions of event phenomenologies, error characteristics, etc. To characterize event behavior over a protracted period of time, it is important that there be consistency in the radar information, which is archived.

The current NEXRAD system is providing important data for a more comprehensive understanding of the climatic variability over the United States. These data include inferences on rainfall, storm intensity, storm tracks, winds, boundary layer changes, urban effects, land-sea interactions, and human interactions. Radar variables in themselves, in contrast to meteorological variables derived from them, can also provide a set of consistent data from which climatic inferences can be made. Furthermore, radar data can be applicable at scales ranging from the microscale to the macroscale. When a replacement radar system is installed, it will be critical to retain continuity in observational content and, as much as possible, to retain formats to ensure preservation of the climatic record for analysis and assessment of changes. The possible relationships of weather variability associated with global climate changes such as temperature and sea state changes are of particular importance, and the availability of a consistent weather radar database within this context will be critical to making and/or validating associated assessments.

Mesoscale radar climatology has been helpful in advancing our understanding of mesoscale convective systems (e.g., Bluestein and Jain, 1985; Houze et al., 1990; Parker and Johnson, 2000), a key factor in the development of convective nowcast products. A practical application is the use of counts and characterizations of microbursts, based on radar observations at several airports, as an input for cost-benefit analyses to support Federal Aviation Administration (FAA) decisions regarding the deployment of wind-shear detection systems (McCarthy and Serafin, 1984).

Recommendation

To support the use of radar data in the climate observing system and other research areas, standards for calibration and continuity of observations should be established and implemented.

DATA ARCHIVING AND ANALYSIS

There are several reasons for the support of a complete and accessible radar data archive:

- support for radar climatology,
- support for research and testing for improved data quality analysis (DQA) and product algorithms, and
- quantification of the data error characteristics, required for hydrology and NWP models.

Climatological and off-line product development involves working with data from an archive. The official NEXRAD archive is maintained at the National Climatic Data Center (NCDC). The archived data must be used in the state in which they were archived, including the selected scan strategy, resolution, and format. The future system should maintain continuity in observational content and, as much as possible, formats in order to ensure preservation of the climatic record for analysis and assessment of changes.

It is unreasonable to attempt to anticipate all uses for these data. However, it is important to keep the data in their most basic form and to let the researchers apply data quality analysis. A desire for consistency also suggests that the data archive be subjected to some core standards. On the one hand, data consistency may facilitate the accurate estimation of data errors. On the other hand, such standards may be at odds with adaptive scanning strategies. "Do no harm" should be an imperative. For example, care must be taken that a faulty control algorithm not be able to deny to the archive data that are essential for the improvement of that algorithm. These considerations are far-reaching.

Recommendation

Plans for next generation weather radar systems should include provisions for real-time dissemination of data to support forecast, nowcast, and warning operations and data assimilation for numerical weather prediction, and certain research applications. Routine reliable data archiving for all radars in the system for research, climatological studies, and retrospective system evaluation must be an integral part of the system. Convenient, affordable access to the data archives is essential.

SUPPORT TO USER DECISION PROCESSES

Long-term consideration should be given to the development of a multitude of products that describe the atmosphere in new and exciting ways and that have a high degree of impact to end users. End users include farmers and ranchers, those involved in transportation on the ground, on the water, or in the air, and, of course, the general public. Many such products cannot be conceived today, until an advanced system is developed.

Also, many tactical decision aids can be generated to change how the user community looks at weather science capability. For example, better weather information would provide the means for traveling from one place to another with much greater efficiency, or the means for maximizing crop yields.

Going beyond this concept is a fully integrated observing system (IOS), which will use radar as but one of many observational capabilities, so that many years hence, a four-dimensional data system (space and time) will be used to characterize all manner of weather situations. The IOS, which may in fact be a number of such systems that address the needs of a variety of users (including many with limited meteorological knowledge), will provide the ability to extract weather features critical to users' needs. User needs include information on severe weather, wind and moisture fields, or on weather as it may impact aviation. A current aviation-specific example illustrates such a capability. With the concept of collaborative decision making—whereby the three elements of the aviation system (the pilot, the air traffic controller, and the airline dispatcher) work with the same weather data system (likely the same data source but displayed in different user-specific ways)—a safer and more efficient utilization of the airspace is achieved.

The end state user may provide input to two types of decision aids: (1) highly directed aids for meteorological users that provide forecasters with improved assistance in operational warning and forecast needs and (2) a potpourri of nonmeteorological products and decision aids for nonmeteorological users such as those in the civil and military aviation sectors, the newly emerging intelligent surface transportation systems (e.g., smart cars and trucks), agriculture, and U.S. military. A user group should be formed in parallel to the evolution of the IOS to ensure effective development of an advanced product suite.

Recommendation

Tactical Decision Aids and means for collaborative decision-making capabilities should be developed for both meteorological and nonmeteorological users of the system, with attention to the demands on the integrated observing system.

6

Findings and Recommendations

As a consequence of the development of the science of radar meteorology and the advent of the NEXRAD system, weather radar has become the primary means of detecting, describing, tracking, and nowcasting or short-term forecasting precipitation-laden and, to a more limited extent, clear air weather over the contiguous United States. It has value both as a stand-alone system and as a major component of the entire atmospheric observing system. The NEXRAD system contributes to the preservation of life and property (e.g., through significantly improved tornado, severe storm, flash flood, and hurricane landfall warnings); to public safety; to the advancement of meteorology, hydrology and climatology; to aviation; and to the conduct of numerous weather-dependent activities. Even though the system's current configuration provides outstanding public service, innovations and technological advances will add significant value to the evolving NEXRAD system and ultimately to the future weather radar system. This future system will most likely include a variety of different sensor types that will ensure continuation of present capabilities, address current deficiencies, and take advantage of opportunities afforded by new knowledge.

The more important uses of weather radar data in the future will be within the context of an integrated observing system. Radar data will be combined with data from satellites, surface networks, commercial aircraft, and other sources to produce weather depictions that are more effective than those in place today. Radar data will also be used increasingly in advanced data assimilation schemes with numerical weather and associated hydrologic prediction models. These applications will require real-time data of very high quality, with error statistics that are well quantified.

The committee was neither constituted nor charged to design the future generation weather surveillance radar system. The findings and recommendations summarized herein deal with technologies having the potential to mitigate some of the limitations of the evolving NEXRAD system. For many of the approaches, and certainly for those categorized as "far-term" or "visionary," technical feasibility remains to be established. In most cases benefit-cost evaluations will have to precede any move toward implementation in the design of the future system. The committee encourages the agencies that commissioned this study to follow through with the investigations necessary to establish the technical feasibility of the "far-term" and "visionary" technologies and to conduct benefit-cost analyses of the feasible ones.

The success of the findings and recommendations summarized herein will depend critically on the development of a parallel end-state user process (i.e., extending beyond the hydrologist or meteorologist) that defines needs in user contexts. The committee believes that without this user check and balance system, the best plans for the future could produce scientific and technical value, but could be of limited user value. Likewise, the concept of computer augmentation and decision support is fundamental to this end process. The complexities of an integrated observing system, and the four-dimensional sorting thereof for specific user needs, may require sophisticated routines that provide for decision support systems.

GROUP I: RADAR TECHNOLOGIES

The findings and recommendations of this committee are summarized here in two groups. The first group concerns those technical approaches considered most promising for the development of future weather radar systems, and is presented in order of our estimate of their maturity (starting with the most mature) of the various approaches.

Finding

The current NEXRAD (WSR-88D) system, a highly capable weather surveillance radar, has proved to be of great value to many sectors of our society, and its applications have extended beyond the traditional goal of protecting life and property. The Radar Operation Center and the NEXRAD Product Improvement Program mechanisms have provided an evolutionary process for improvements to the system.

Recommendation—Near-term (Chapter 2)

The Radar Operation Center and the NEXRAD Product Improvement Program mechanisms should be extended to permit continual improvement to the NEXRAD system. Provisions should be made to carry features found to be beneficial, such as polarization diversity, over to the succeeding generation of systems.

Finding
Other radar systems provide various kinds of weather surveillance. Examples include the Terminal Doppler Weather Radar (TDWR) and Airport Surveillance Radars (ASR) operated by the Federal Aviation Administration (FAA), radars operated by commercial broadcasters and other private entities, clear air profilers, and airborne weather avoidance radars carried on commercial and private aircraft.

Recommendation—Near-term (Chapter 4)

The potential value and technology to incorporate data from complementary radar systems to provide a more comprehensive description of the atmosphere should be investigated

Finding
Mobile radars can provide highly detailed views of weather events. Such observations not only have scientific interest, but also could be valuable in support of emergency services in cases such as fires, contaminant releases, and nuclear, chemical, or biological attacks.

Recommendation—Near-term (Chapter 4)

The potential of operational mobile radar systems to contribute to the nation's weather surveillance system for emergency response and for improved short-term forecasts should be evaluated.

Finding
Adaptive waveform selection and volume scan patterns are important for optimizing radar performance in different weather situations. Beam and waveform management will be determined by the prevailing atmospheric phenomena and threats.

Recommendation—Far-term (Chapter 3)

Adaptive waveform selection, which may even be applied to present systems, and agile beam scanning strategies, which require an electronically scanned phased array system, should be explored to optimize performance in diverse weather.

Finding
Radar systems with phased array antennas and advanced waveforms can support a broad spectrum of applications with observation times sufficiently short to deal with rapidly evolving weather events such as tornadoes or downburst winds.

Recommendation—Far-term (Chapter 3)

The technical characteristics, design, and costs of phased array radar systems that would provide the needed rapid scanning, while preserving important capabilities such as polarization diversity, should be established.

Finding

A closely spaced network of short-range radar systems would provide near-surface coverage over a much wider area than the current NEXRAD system. This network would expand geographic coverage of low-level winds, precipitation near the surface, and weather phenomena in mountainous regions.

Recommendation—Far-term (Chapter 4)

The potential for a network of short-range radar systems to provide enhanced near-surface coverage and supplement (or perhaps replace) a NEXRAD-like network of primary radar installations should be evaluated thoroughly.

Finding

The satellite-borne Tropical Rainfall Measurement Mission (TRMM) radar has demonstrated the ability to measure precipitation over regions not reached by land-based radars. The TRMM data are useful for climate monitoring and extended-range weather and climate forecasting. Future satellite technology is likely to allow on-orbit operation of weather radar systems with larger antenna apertures and higher power outputs than are currently used in space. Satellite constellations operating as distributed array antennas would provide high-resolution global coverage, thereby supplementing ground-based networks.

Recommendation—Visionary (Chapter 4)

The capabilities of future space-based radar systems to supplement ground-based systems should be determined.

Finding

Both piloted and Unmanned Aerospace Vehicles (UAV) are being developed for a variety of remote sensing and other applications. As the capabilities of these airborne platforms increase it may become possible to place weather radar systems on station at a variety of altitudes, for extended duration. These systems could support weather forecasting, weather warnings and other emergency response applications.

Recommendation—Visionary (Chapter 4)

The capabilities of Unmanned Aerospace Vehicles and piloted aircraft to carry weather radar payloads should be monitored for their potential to provide weather surveillance over the continental United States and over the oceans.

GROUP II: PROCEDURES

This second group of findings and recommendations concerns matters important to the design, development, and implementation of future generations of weather surveillance radar systems, independent of the specific technologies that may be adopted. No order is intended for the following recommendations.

Finding

Weather forecasting and warning applications are increasingly relying on integrated observations from a variety of systems that are asynchronous in time and are nonuniformly spaced geographically. Weather radar is a key instrument that provides rapid update and volumetric coverage.

Recommendation (Chapter 1)

The next generation of radars should be designed as part of an integrated observing system aimed at improving forecasts and warnings on relevant time and space scales.

Finding

Weather surveillance needs vary from region to region and from season to season, depending on factors such as the depth of precipitating cloud systems and local topography.

Recommendation (Chapter 2)

Weather surveillance needs should be evaluated by geographic region to determine if a common radar system design is appropriate for all regions.

Finding

Early field-testing of NEXRAD concepts and systems in a limited range of geographic and climatological situations did not evaluate and elucidate the full range of operational demands on the system.

Recommendation (Chapter 2)

The development program for the next generation weather surveillance radar system should incorporate adequate provision for beta testing in the field in locations with diverse climatological and geographic situations.

Finding

Despite the great utility and value of NEXRAD, data-quality issues, addressed in part through the NEXRAD Product Improvement Program (NPI), remain a significant impediment to some important applications.

Recommendation (Chapter 3)

The quality of real-time data should receive prominent consideration in the design and development of a next generation weather surveillance radar system. Real-time data quality assessment should be automated and used in deriving error statistics, and alerting users to system performance degradation.

Finding

Present and prospective users of the electromagnetic spectrum are competing for the current weather radar spectrum allocation. There is particular concern that the use of S-band may be lost for weather radar applications. S-band minimizes attenuation effects and ambiguities in Doppler wind measurements. The loss of S-band would compromise measurements of heavy rain and hail, warnings of flash floods and tornadoes, and the monitoring of hurricanes near landfall. The cost of rectifying these impacts in the current NEXRAD system would be substantial. Transmitters, receivers, and software would all require replacement and/ or significant modifications. It is also possible that antennas would need to be replaced. The current state of science is such that certain capabilities would be lost.

Recommendation (Chapter 3)

Policy makers and members of the operational community should actively participate in the arena of frequency allocation negotiation. The impact, including the economic and societal costs, of restrictions on operating frequency, bandwidth, and power should be assessed for current and future weather radar systems.

Finding

Weather radar data are being increasingly used in climatological studies as well as for a wide variety of other research. Weather radars provide continuous high-resolution monitoring in space and time.

Recommendation (Chapter 5)

To support the use of radar data in the climate observing system and other research areas, standards for calibration and continuity of observations should be established and implemented.

Finding

Weather radar provides observations, on small spatial and time scales, that are essential for monitoring precipitation and diagnosing certain weather events as well as for supporting nowcasting systems, hydrologic models, and numerical weather prediction models. Effective assimilation of radar data in these models requires accurate error statistics.

Recommendation (Chapter 5)

The value of radar data, as part of an integrated observing system, in diagnostic applications nowcasting systems, and hydrologic and numerical weather prediction models should be considered in the design of the next generation weather radar system. The characteristics of radar observations and associated error statistics must be quantified in ways that are compatible with user community needs.

Finding

Broad dissemination of weather radar data in real time facilitates the application of those data to diagnostic and forecasting operations. Archiving of radar base data, as well as product data, facilitates research activities, retrospective studies, and climatological investigations.

Recommendation (Chapter 5)

Plans for next generation weather radar systems should include provisions for real-time dissemination of data to support forecast, nowcast, and warning operations and data assimilation for numerical weather prediction, and certain research applications. Routine reliable data archiving for all radars in the system for research, climatological studies, and retrospective system evaluation must be an integral part of the system. Convenient, affordable access to the data archives is essential.

Finding

A long-term objective of the radar and other weather observation systems mentioned in this report is the establishment of an integrated observational system, whereby most or all of these observations (e.g., ground-based and space-borne radar; hyperspectral, visual, and IR satellite data; and directly and remotely sensed data from manned and unmanned aircraft, commercial aircraft, surface-based remote sensors, and radiosondes), would be assimilated on to a fixed four-dimensional grid to provide the most complete diagnosis of weather impacts possible. Additionally, numerical weather prediction (NWP) models and now-casting techniques would readily provide forecasts for times ranging from a few minutes to many hours. A broad array of products will be used to support decisions that will improve safety to humans, improve operational efficiency, and make homeland defense efforts more effective.

Recommendation (Chapter 5)

Tactical Decision Aids and means for collaborative decision-making capabilities should be developed for both meteorological and nonmeteorological users of the system, with attention to the demands on the integrated observing system.

7

Concluding Remarks:
Radar in a Time of Terrorism

This study began during spring 2001, a time marked by unprecedented peace and prosperity in the United States. On September 11, 2001 horrific acts of terrorism rocked the foundation of this country. In subsequent weeks, fears of further domestic terrorism continued, and fears of potential atmospheric releases of chemical, biological, or nuclear materials increased. This report has been prepared in this context. Though a detailed investigation of the application of weather radar to such situations is beyond the scope of this study, the committee feels there are important ways in which weather radar is directly tied to homeland security, and this need should be raised in the context of this report.

The current radar system could be employed in support of tactical and strategic aids for characterizing the transport and deposition of contaminants near the earth's surface. This could be accomplished by: (1) making direct use of NEXRAD Doppler wind data, (2) assimilating the Doppler wind data into high-resolution mesoscale models to forecast transport and deposition characteristics, (3) deploying emergency relocatable radars similar to those used in tornado research to regions of expected or actual terrorist releases of contaminants, and (4) characterizing and quantifying scavenging and deposition rates of dangerous materials by precipitation (e.g., Seliga et al., 1989; Jylhä, 1999a,b). Some of these possibilities are to be tested in an urban dispersion experiment planned for the Oklahoma City area in 2003.

Future generation weather surveillance radar systems could provide similar, but improved, capabilities in this important area. Because of the potentially important use of weather surveillance radar as a tool in homeland security, it is critical that these discussions and investigations continue.

References

Anagnostou, E. N., W. F. Krajewski, D. J. Seo, and E. R. Johnson. 1998. Mean-field radar rainfall bias studies for WSR-88D. ASCE Journal of Engineering Hydrology 3(3):149–159.

Atlas, D. 1990. Radar in Meteorology. Boston: American Meteorological Society.

Atlas, D. and R. K. Moore. 1987. The measurement of precipitation with Synthetic Aperture Radar. J. Atmos. Ocean. Tech. 4:368–376.

Atlas, D., C. W. Ulbrich, F. D. Marks, Jr., E. Amitai, and C. R. Williams. 1999. Systematic variation of drop size and radar-rainfall relations. J. Geophys. Res. 104(D6):6155–6169.

Atlas, R., R. N. Hoffmanb, S. M. Leidner, J. Sienkiewicz, T. W. Yu, S. C. Bloom, E. Brin, J. Ardizzone, J. Terry, D. Bungato, and J. C. Bloom. 2001. The effect of marine winds from scatterometer data on weather analysis and forecasting. B. Am. Meteorol. Soc. 82.

Baldwin, M. P. and T. J. Dunkerton. 2001. Stratospheric harbingers of anomalous weather regimes. Science 294:581–584.

Balsey, B. B., and K. S. Gage. 1980. The MST radar technique: Potential for middle atmospheric studies. Pure and Applied Geophysics 118:452–493.

Barton, D. K., C. E. Cook and P. Hamilton, eds. 1991. Radar Evaluation Handbook. Boston: Artec House.

Battan, L. J. 1973. Radar observation of the atmosphere. Chicago: Univ. of Chicago Press.

Beguin, D. and J. L. Plante. 1998. Critical technology requested by fast scanning radar. COST 75 Final International Seminar on Advanced Weather Radar Systems, Locarno, Switzerland, 645–657.

Billam, E. R. and D. H. Harvey. 1987. MESAR—An advanced experimental phased array radar. Proceedings of the IEEE International Radar Conference, London, 37–40.

Bluestein, H. B. and M. H. Jain. 1985. Formation of mesoscale lines of precipitation source squall lines in Oklahoma during the spring. J. Atmos. Sci. 42:1711–1732.

Bluestein, H. B. and A. L. Pazmany. 2000. Observations of tornadoes and other convective phenomena with a mobile, 3-mm wavelength, Doppler radar: The spring 1999 field experiment. B. Am. Meteorol. Soc. 81(2):939–951.

Briggs, B. 1980. Radar observation of atmospheric winds and turbulence: A comparison of techniques. J. Atmos. Terr. Phys. 42:823–833.

Bringi, V. N. and V. Chandrasekar. 2001. Polarimetric Doppler Weather Radar: Principles and Applications. Cambridge, New York: Cambridge University Press.

Bringi, V. N., T. A. Seliga and K. Aydin. 1986. Hail detection with a differential reflectivity radar. Science 225:1145–1147.

Bucci, N. J., and H. Urkowitz. 1993. Testing of Doppler tolerant range sidelobe suppression in pulse compression meteorological radar. Proceedings of the IEEE National Radar Conference, Boston, 206–211.

Carbone, R. E., J. D. Tuttle, D. Ahijevych, S. B. Trier, and C. A. Davis. 2002. Inferences of predictability associated with warm season precipitation episodes. J. Atmos. Sci. 59(13):2033–2056.

Carey, L. D., and S. A. Rutledge. 1998. Electrical and multiparameter radar observations of a severe hailstorm. J. Geophys. Res. 103:13979–14000.

Caylor, I. J. and V. Chandrasekar. 1996. Time Varying Ice Crystal Orientation Observed with Multiparameter Radar. IEEE T. Geosci. Remote 34(4):847–858.

Chandrasekar, V., and A. P. Jayasumana. 2001. Radar Design and Management in a Networked Environment. P. ITCOMM, Denver CO, 142–147.

Chandrasekar, V., V. N. Balakrishnan and D. S. Zrnic. 1990. Error structure of multiparameter radar and surface measurements of rainfall. Part III: Specific differential phase. J. Technol. 7:621–629.

Chandrasekar, V., D. Brunkow and A. P. Jayasumana. 2001. CSU-CHILL Operation over the Internet: Virtual CSU–CHILL. Preprints, 30th International Conference on Radar Meterology, AMS, Boston, 58–60.

Chandrasekar ,V, R. Meneghini and I. I. Zawadzki. 2002. Local and global measurement of precipitation. Atlas Symposium Conference Volume, AMS, Boston.

Ciach, J. G. and W. F. Krajewski. 1999. On the estimation of radar rainfall error variance. Advances in Water Resources 22(6):585–595.

Collier, C. G., ed. 2001. COST-75: Advanced weather radar systems 1993–1997 Final Report. Brussels: Office of Publication of European Communities.

Crum, T. D., and R. L. Alberty. 1993. The WSR-88D and the WSR-88D operational support facility. B. Am. Meteorol. Soc. 74:1669–1687.

Crum, T. D., R. L. Alberty and D. W. Burgess. 1993. Recording, archiving, and using WSR-88D data. B. Am. Meteorol. Soc. 74:645–653.

Dobaie, A. and V. Chandrasekar. 1995. Application of Tomographic Technique for Analysis of Differential Propagation Phase Measurements. Preprints, 27th Conference on Radar Meteorology, AMS, Boston 389–391.

Doviak R. J. and R. S. Zrnic. 1993. Doppler Radar and Weather Observations. San Diego: Academic Press.

Droegemeier, K. K., K. Kelleher, T. Crum, J. J. Levit, S. A. Del Greco, L. Miller, C. Sinclair, M. Benner, D. W. Fulker, and H. Edmon. 2002. Project CRAFT: A test bed for demonstrating the real time acquisition and archival of WSR-88D Level II data. Preprints, 18th International Conference on Interactive Information Processing Systems (IIPS) for Meteorology, Oceanography, and Hydrology, 13–17 January, AMS, Orlando, Fla. 136–139.

Droegemeier, K. K., J Smith, S. Businger, C. Doswell III, J. Doyle, C. Duffy, E. Foufoula-Georgiou, T. Graziano, L. D. James, V. Krajewski, M. LeMone, D. Lettenmaier, C. Mass, R. Pielke, Sr., P. Ray, S. Rutledge, J. Schaake, and E. Zipser. 2000. Hydrological aspects of weather prediction and flood warnings: Report of the ninth prospectus development team of the USWRP. B. Am. Meterol. Soc. 81, 2665–2680

Evans, J. E., and E. R. Ducot. 1994. The integrated terminal weather system (ITWS). Lincoln Laboratory Journal 7(2):449–474.

Fabry, F., C. Frush, I. Zawadzki, and A. Kilambi. 1997. On the Extraction of near-surface index refraction using radar phase measurements from ground targets. J. Atmos. Ocean. Tech. 14:978–987.

Facundo, J. 2000. Update on Commissioning the Advanced Weather Interactive Processing System (AWIPS). 16th International Conference on Interactive Information and Processing Systems for Meteorology, Oceanography, and Hydrology, AMS, Long Beach, 291–294.

Falkovich, A., E. Kalnay, S. Lord, M. B. Mathur. 2000. A new method of observed rainfall assimilation in forecast models. J. Appl. Meteorol. 39:1282–1298.

Fitch, J. P. 1988. Synthetic Aperture Radar. New York: Springer Verlag.

Forsyth, D. E., K. J. Kimpel, D. S. Zrnic, S. Sandgathe, R. Ferek, J. F. Heimmer, T. McNellis, J. E. Crain, A. M. Shapiro, J. D. Belville, and W. Benner. 2002. The National Weather Radar Testbed (Phased Array). 18th International Conference on Interactive Information and Processing Systems (IIPS) AMS, Orlando, Fla.

Fritsch, J.M., R. A. Houze Jr., R. Adler, H. Bluestein, L. Bosart, J. Brown, F. Carr, C. Davis, R. H. Johnson, N. Junker, Y. H. Kuo, S. Rutledge, J. Smith, Z. Toth, J. W. Wilson, E. Zipser, and D. Zrnic. 1998. Quantitative precipitation forecasting: Report of the eighth prospectus development team, U.S. Weather Research Program, B. Am. Meteorol. Soc. 79:2.

Gossard, E. E. 1990. Radar research on the atmospheric boundary layer. Pp. 477–527 in Radar in Meteorology, D. Atlas, ed. Boston: American Meteorological Society.

Grecu, M. and W. F. Krajewski. 2000. Effect of model uncertainty in variational assimilation of radar data on rainfall forecasting. J. Hydrol. 239(1–4):85–96.

Haase G., P. Gross and C. Simmer. 1999. Physical initialisation of the Lokalmodell (LM) using radar data. Third International SRNWP Workshop on Non-Hydrostatic Modelling, Offenbach, Germany.

Hansen, R. C. 1998. Phased Array Antennas. New York: John Wiley.

Holloway, C. L. and R. J. Keeler. 1993. Rapid Scan Doppler Radar: The antenna issues, 26th Conference on Radar Meteorology, AMS, Boston, 393–395.

Houze, R. A., B. F. Smull, and P. Dodge. 1990. Mesoscale organization of springtime rainstorms in Oklahoma. Mon. Weather Rev. 118:613–654.

Institute of Electrical and Electronics Engineers. 1990. IEEE Standard Radar Definitions. IEEE Standard No. 686–1990.

Josefsson, L. 1991. Phased array antenna technology for weather radar applications. Preprints, 25th Conference on Radar Meteorology, AMS, Paris, 752–755.

Jylhä, K. 1999a. Relationship between the scavenging coefficient for pollutants in precipitation and the radar reflectivity factor. Part I: Derivation. J. Appl. Meteorol. 38:1421–1434.

Jylhä, K. 1999b. Relationship between the scavenging coefficient for pollutants in precipitation and the radar reflectivity factor. Part II: Applications. J. Appl. Meteorol. 38:1435–1447.

Jylhä, K. 2000. Assessing Precipitation Scavenging of Air Pollutants by Using Weather Radar. Phys. Chem. Earth Pt. B 25:1085–1089.

Katz, S. L., B. Hudson, H. Pschunder and J. D. Lupnacca. 1997. Aircraft Wake Vortex Detection Trials With A C-Band Instrumentation Radar. 28th International Conference on Radar Meteorology, Austin, TX.

Keeler, R. J., and S. M. Ellis. 2000. Observational error covariance matrices for radar data assimilation. Phys. Chem. Earth 25:10–12, 1277–1280.

Keeler, R. J. and C. Frush. 1983. Rapid Scan Doppler Radar Development and Considerations: Technology Assessment. 21st Conference on Radar Meterology, AMS, Boston, 284–290.

Keeler, R. J. and L. J. Griffiths. 1978. Acoustic Doppler extraction by adaptive linear prediction filtering. J. Acoust. Soc. Am. 61:1218–1227.

Keeler, R. J. and C. A. Hwang. 1995. Pulse compression for weather radar. IEEE International Radar Conference, Washington, D.C., 529–535.

Keeler, R. J. and R. E. Passarelli. 1990. Signal processing for atmospheric radars. Pp. 199–229 in Radar in Meteorology, D. Atlas, ed. Boston: American Meteorological Society.

Keenen, T. J., J. Wilson, P. Joe, C. Collier, B. Golding, D. Burgess, R. Carbone, A. Seed, P. May, L. Berry, J. Bally, and C. Pierce. 2002 (in press). The World Weather Research Program (WWRP) Sydney 2000 Forecast Demonstration Project: Overview. B. Am. Meteorol. Soc. 83.

Klazura, G. E., and D. A. Imy. 1993. A description of the initial set of analysis products available from the NEXRAD WSR-88D system. B. Am. Meterol. Soc. 74:1293–1311.

Koivunen, A. C. and A. B. Kostinski. 1999. Feasibility of data whitening to improve performance of weather radar. J. Appl. Meteorol. 38:741–749.

Krishnamurti, H. S. Bedi, K. S. Yap, and D. Oosterhof. 1993. Hurricane forecasts in the FSU models. Adv. Atmos. Sci. 10:121–132.

Kuo, Y. H., S. Sokolovskiy, R. Anthes, and F. Vandenberghe. 2000. Assimilation of GPS radio occultation data for numerical weather prediction. Terr. Atmos. Ocean Sci. 11(1):157–186.

Liu, H., V. Chandrasekar. 2000. Classification of Hydrometeors Based on Polarimetric Radar Measurements: Development of Fuzzy Logic and Neuro-Fuzzy Systems, and In Situ Verification. J. Atmos. Ocean. Tech. 17(2):140–164.

Maese, T., J. Melody, S. Katz, M. Olster, W. Sabin, A. Freedman, and H. Owen. 2001. Dual-use shipborne phased array radar technology and tactical environmental sensing. Proceedings of the IEEE National Radar Conference, Atlanta, 7–12.

Martner, B. E., K. A. Clark, S. Y. Matrosov, W. C. Campbell, and J. S. Gibson. 2001. NOAA/ETL's polarization-upgraded X-band "Hydro" radar. Preprints, 30th International Conference on Radar Meteorology, Munich.

Martner, B. E., D. B. Wuertz, B. B. Stankov, R. G. Strauch, E. R. Westwater, K. S. Gage, W. L. Ecklund, C. L. Martin, and W. F. Dabbert. 1993. An evaluation of wind profiler RASS, and microwave radiometer performance. B. Am. Meteorol. Soc. 74:599–613.

McCarthy, J., and R. Serafin. 1984. The microburst: Hazard to aircraft. Weatherwise 37(3):120–127

Meischner, P., C. Collier, A. Illingworth, J. Joss, and W. Randeu. 1997. Advanced Weather Radar Systems in Europe: The COST 75 Action. B. Am. Meteorol. Soc., 78:1411–1430.

Morris, D. A., K. C. Crawford, K. A. Kloesel, and J. M. Wolfinbarger. 2001. OK-FIRST: A meteorological information system for public safety. B. Am. Meteorol. Soc. 82:1911–1923.

Mudukutore, A., V. Chandrasekar, and R. J. Keeler. 1998. Pulse Compression for Weather Radars: Simulation and Evaluation. IEEE T. Geosci. Remote. 36(1):125–142.

Neiman, P. J., P. T. May, and M. A. Shapiro. 1992. Radio Acoustic Sounding System (RASS) and wind profiler observations of lower- and mid-tropospheric weather systems. Mon. Weather Rev. 129:2298–2313.

NRC. 1995. Assessment of NEXRAD Coverage and Associated Weather Services. Washington, DC: National Academy Press.

NRC. 1999. A Vision for the National Weather Service Roadmap for the Future. Washington, DC: National Academy Press.

Palmer, R. D., T. Y. Yu and P. B Chilson. 1999. Range imaging using frequency diversity, Radio Sci. 34(6):1485–1496.

Palmer R. D., S. Gopalam, T. Y. Yu, and S. Fukao. 1998. Coherent imaging using Capon's method. Radio Sci. 33(6):1585–1598.

Parker, M. D. and R. H. Johnson. 2000. Organizational modes of mid-latitude mesoscale convective systems, Mon. Weather Rev. 128:3413–3436.

Protopapa, A., C. Boni, M. Di Lazzaro, M. De Fazio, F. A. Studer, N.J. Bucci, S. L. Katz, H. Urkowitz, J. J. Gallagher and J. D. Nespor. 1994. A description of a multifunction phased array primary radar for terminal area surveillance. COST-75 Weather Radar Systems International Seminar, Brussels, 650.

Radar Operations Center. 2002. Interface Control Document for the RPG to Class 1 User. Document 2620001C. Norman, OK: Radar Operations Center.

Rich, S. T. 1992. Integrating wind profiler data into forecast and warning operations at NWS field offices. NOAA Tech. Memo. NWS SR-141, 34. Boulder, Colo.: NOAA Forecast Systems Lab.

Richardson, Y. P., D. Dowell, J. Wurman, P. Zhang, and S. Weygandt. 2001. Preliminary dual-Doppler analyses of two tornadic thunderstorms. Preprints, 30th International Conference on Radar Meteorology, Munich, 295–297.

Rocken, C., R. Anthes, M. Exner, D. Hunt, S. Sokolovskiy, R. Ware, M. Gorbunov, W. Schreiner, D. Feng, B. Herman, Y. Kuo, and X. Zou. 1997. Analysis and validation of GPS/MET data in the neutral atmosphere. J. Geophys. Res. 102(D25):29849–29866.

Rogers, J. W., L. Buckler, A. C. Harris, M. Keehan, C. J. Tidwell. 1997. History of the Terminal Area Surveillance System (TASS). Preprints, 28th Conference on Radar Meterology, AMS, Austin, TX 157–158.

Rogers, R. R. and P. L. Smith. 1996. A short history of radar meteorology. In Historical Essays on Meteorology 1919–1995, J.R. Fleming, ed. Boston: American Meteorological Society.

Sachidananda, M. and D. S. Zrnic. 1986. Differential propagation phase shift and rainfall rate estimation. Radio Sci. 21:235–247.

Sachidananda, M and D. S. Zrnic. 1987. Rain rate Estimation from Differential Polarization Measurements. J.Atmos. Ocean. Tech. 4:588–598.

Saffle, R. E., M. I. Istok and L. D. Johnson. 2001. NEXRAD Open Systems—Progress and Plans. Preprints, 30th International Conference on Radar Meteorology, AMS, Munich 690–693.

Sauvageot, H. 1992. Radar Meteorology. Boston: Artech House.

Scarchilli, G., E. Gorgucci and V. Chandrasekar, 1995: Self Consistency of Polarization Diversity Measurements of Rainfall. IEEE T. Geosci. Remote. 34(1):22–26.

Schuur, T. J., D. S. Zrni_ and R. E. Saffle. 2001. The Joint Polarization Experiment—An operational test of weather radar polarimetry. Preprints, 30th International Conference on Radar Meteorology, AMS, Munich, 722–723.

Seliga, T. A. and V. N. Bringi. 1976. Potential use of radar differential reflectivity measurements at orthogonal polarizations for measuring precipitation. J. Appl. Meteorol. 15:69–76.

Seliga, T. A., and V. N.Bringi. 1978. Differential reflectivity and differential phase shift: Applications in radar meteorology. Radio Sci. 13: 271–275.

Seliga, T. A., H. Direskeneli and K. Aydin. 1989. Potential role of differential reflectivity to estimate scavenging of aerosols, Preprints, 24th Conference on Radar Meteorology, AMS, Tallahassee, Fla., 363–366.

Seminario, M., K. Gojara and V. Chandrasekar. 2001. Noise correction of polarimetric radar measurements. Preprints, 30th International Conference on Radar Meteorology, AMS, Boston, 38–40.

Seo, D. J., J. P. Breidenbach, R. Fulton, D. Miller and T. O'Banon. 2000. Real time adjustment of range dependent biases in WSR-88D rainfall estimates due to nonuniform vertical profile of reflectivity. J. Hydrometeorol. 1:222–240.

Serafin, R. J. and J. W. Wilson. 2000. Operational weather radar in the United States: Progress and opportunity. B. Am Meteorol. Soc. 81:501–518.

Serafin, R. J. 1996. The evolution of atmospheric measurement systems. In Historical Essays on Meteorology 1919–1995, J. R. Fleming, ed. Boston: American Meteorological Society.

Skolnik, M. I. 2001. Introduction to Radar Systems. New York: McGraw-Hill.

Smith, J. A., D. J. Seo, M. L. Baeck, and M. D. Hudlow. 1996. An intercomparison study of NEXRAD precipitation estimates. Water Resour. Res. 32:2035–2045.

Smith, P. L. 1995. Dwell-time considerations for weather radars. Preprints, 27th Conference on Radar Meteorology, AMS, Vail, Colo. 760–762.

Smith, P.L. 1974. Applications of radar to meteorological operations and research. Pr. Inst. Electr. Elect. 62:724–745.

Srivastava, R. C., and L. Tian. 1996. Measurement of attenuation by a dual-radar method: Concept and error analysis. J. Atmos. Ocean. Tech. 13:937–947.

Sun, J. and N. A. Crook. 1997a. Dynamical and microphysical retrieval from Doppler radar observations using a cloud model and its adjoint. Part I: Model development and simulated data experiments. J. Atmos. Sci. 54:1642–1661.

Sun, J., and N. A. Crook. 1997b. Dynamical and microphysical retrieval from Doppler radar observations using a cloud model and its adjoint. Part II: Retrieval experiments of an observed Florida convective storm. J. Atmos. Sci. 55:835–852.

Testud, J. and P. Amayenc. 1989. Stereo-radar meteorology: A promising method to observe precipitation from a mobile platform. J. Atmos. Ocean. Tech. 6:89–108.

Tokay, A., D. A. Short, C. R. Williams, W. L. Ecklund, and K .S. Gage. 1999. Tropical rainfall associated with convective and stratiform clouds: Intercomparison of disdrometer and profiler measurements. J. Appl. Meteorol. 38:302–320

Torres, S. M. and D. S. Zrnic. 2001. Optimum processing in range to improve estimates of Doppler and Polarimetric Variables. Preprints, 30th International Conference on Radar Meteorology, AMS 325–327.

Trapp, R. J., E. D. Mitchell, G. A. Tipton, D. W. Effertz, A. I. Watson, D. L. Andra, and M. A. Magsig. 1999. Descending and nondescending tornadic vortex signatures detected by WSR-88Ds. Weather Forecast. 14:625–639.

Treadon, R. E. 1996. Physical initialization in the NMC global data assimilation system. Meteorol. Atmos. Phys. 60:5786.

U.S. Weather Research Program. 2001. An Implementation Plan for Research in Quantitative Precipitation Forecasting and Data Assimilation. Boulder, CO: U.S. Weather Research Program.

Vivekanandan, J., D. S. Zrnic, S. M. Ellis, R. Oye, A. V. Ryzhkov, and J. Straka. 1999. Cloud microphysics retrieval using S-band dual polarization radar measurements. B. Am. Meteorol. Soc. 80:381–388.

Wakimoto, R. M., and C. H. Liu. 1998. The Garden City, Kansas storm during VORTEX 95. Part II: The wall cloud and tornado. Mon. Weather Rev. 126:393–408.

Ware, R. H., D. W. Fulker, S. A. Stein, D. N. Anderson, S. A. Avery, R. D. Clark, K. K. Droegemier, J. P. Kuettner, J. P. Minster, and S. Sorooshian. 2000. SuomiNet: A real-time national network for atmospheric research and education. B. Am. Meteorol. Soc. 81:677–694.

Westrick, K. J., C. F. Mass and B. A. Colle. 1999. The limitations of the WSR-88D radar network for quantitative precipitation measurement over the coastal western United States. B. Am. Meterol. Soc. 80:2289–2298.

Whiton, R. C., P. L. Smith, S. G. Bigler, K. E. Wilk, and A. C. Harbuck. 1998. History of operational use of weather radar by U.S. Weather Services. Part I: The Pre-NEXRAD era; Part II: Development of operational Doppler weather radars. Weather Forecast. 13:219–252.

Wilson, J. W., T. M. Weckwerth, J. Vivekanandan, R. M. Wakimoto, and R. W. Russell. 1994. Boundary layer clear-air radar echoes: Origin of echoes and accuracy of derived winds. J. Atmos. Oceanic Technol. 11:1184–1206.

Wurman, J., and S. Gill. 2000. Fine-Scale Radar Observations of the Dimmitt, Texas Tornado. Mon. Weather Rev. 128:2135–2164.

Wurman, J., S. Heckman and D. Boccippio. 1993. A bistatic multiple-Doppler radar network. J. Appl. Meteorol. 32:1802–1814.

Xue M., D. Wang, D. Hou, K. Brewster and K. K. Droegemeier. 1998. Prediction of the 7 May 1995 squall line over the central U.S. with intermittent data assimilation, 12th Conference on Numerical Weather Predicition, AMS, Phoenix, AZ 191–194.

Zhang, J. 1999. Moisture and Diabatic Initializations Based On Radar and Satellite Observations. Ph.D. dissertation, University of Oklahoma.

Appendix A

NEXRAD WSR-88D System Characteristics

Parameter/Feature	Value/Description
Radar System	
Range of observation	
Reflectivity	460 km
Velocity	230 km
Angular Coverage	
Azimuth	Full circle or sector
Elevation	Operational limits; −1° to +20°
Antenna	
Type	S-Band, center-fed, parabolic dish
Reflector aperture	8.54-m (28-ft) diameter; circular
Beamwidth (one-way, 3 dB)	0.96° at 2.7 GHz; 0.88° at 3.0 GHz
Gain	45.8 dB at 2.85 GHz (midband)
Polarization	Linear horizontal
First side-lobe level	−29 dB
Steerability	360° azimuth; −1° to +45° elevation
Mechanical limits	−1° to +60°
Rotation rate	30° s^{-1} (azimuth and elevation)
Angular acceleration	15° s^{-2} (azimuth and elevation)
Pointing accuracy	± 0.2°

Radome
Type   Fiberglass skin foam sandwich
Diameter   11.89 m (39 ft.)
RF Loss (two-way)   0.3 ± 0.06 dB over 2.7–3.0 GHz band

Transmitter
Type   Master Oscillator Power Amplifier (MOPA)
Frequency range   2.7–3.0 GHz
Peak power output (nominal)   500 kW into antenna
Pulsewidth (nominal)   1.57 μs (short pulse); 4.5 μs (long pulse) ± 4%
RF duty cycle (maximum)   0.002

Pulse Repetition Frequency
Long pulse   322–422 Hz ± 1.7%
Short pulse   322–1282 Hz ± 1.7%
Waveform types   Contiguous and batch

Receiver
Type   Linear
Tunability (frequency range)   2.7–3.0 GHz
Bandwidth (3 dB)   0.63 MHz (short pulse); 0.22 MHz (long pulse)
Phase control   Selectable
Receiver channels   Linear output I/Q; log output
Dynamic range   95 dB max; 93 dB at 1 dB compression
Minimum detectable signal   −113 dBm
Noise temperature   450 K
Intermediate frequency   57.6 MHz
Sampling rate   600 kHz

Signal Processor
Type   Hardwired/programmable
Parameters derived   Reflectivity; mean radial velocity; Doppler
                                    spectral width
Algorithms (respective)   Power averaging; pulse-pair; single-lag cor-
                                    relation

Accuracy (Standard Deviation)
Reflectivity   < 1 dB
Velocity and spectrum width   < 1 m s^{-1}

Number of Pulses Averaged
Reflectivity   6–64
Velocity and spectrum width   40–200

Range Resolution
Reflectivity   1 km
Velocity and spectrum width   0.25 km

Azimuth Resolution
Reflectivity   1°
Velocity and spectrum width   1°
Clutter canceller   Digital, infinite impulse response (IIR), 5-pole
Clutter suppression   30–50 dB
Filter notch half-width   0.5–4 m s^{-1}

Radar Product Generator (RPG)
RPG processor   32-bit general purpose digital computer
Shared memory   32-MB semiconductor memory, expandable
                                         to 96 MB
Wide band communication   1.544 Mbits s^{-1} data rate
Narrow band communication   Up to 21 of 9,600/4,800 bits s^{-1} 4-wire
                                         Up to 26 of 9,600/4,800 bits s^{-1} 2-wire
                                         (will have 14,400/9,600/4,800 bits s^{-1} capa-
                                         bility)

RPG Graphic Display Processor
Principal user processor (PUP)   Fixed point: 32-bit general purpose digital
                                         computer
Communications   9600/488 bits s^{-1}; 2- and 4-wire (maximum:
                                         10 lines)
                                         RS449/RS2332 converters
Video   Color, with split-screen and zoom features
Mass storage   Up to two 600 MB disks

Other
National Climatic Data Center   National archive for NEXRAD data and other
(NCDC)   meteorological and climatological data.
Level II archive   Interface located at the Radar Data Acquisi-
                                         tion (RDA). Digital base data output from the
                                         signal processor that includes base reflectivity,
                                         mean radial velocity, and spectrum width.
                                         Other data include information on synchroni-
                                         zation, calibration, date, time, antenna position,
                                         and operational mode. Recorded on 8-mm
                                         magnetic tape and sent to NCDC for perma-
                                         nent storage.

Level III archive	Interface is located at the Radar Product Generator (RPG). A set of pre-determined products defined in FMH-11 Part A. Data are archived on WORM Optical Disk and sent to NCDC for permanent storage.
National Weather Radar Network	Consists of WSR-88D sites dispersed throughout the conterminous United States plus Department of Defense sites or non-CONUS Department of Transportation sites.

APPENDIX B

Acronyms

4DDA	Four-Dimensional Data Assimilation
AFWA	Air Force Weather Agency
AMS	American Meteorological Society
ARSR	Air Route Surveillance Radar
ASR	Airport Surveillance Radars
AWIPS	Advanced Weather Interactive Processing System
BDDS	Base Data Distribution System
BLM	Bureau of Land Management
CDM	Collaborative Decision Making
CIWS	Corridor Integrated Weather System
CODE	Common Operations and Development Environment
CONUS	Continental United States
COSMIC	Constellation Observing System for Meteorology, Ionosphere, and Climate
CRAFT	Collaborative Radar Acquisition Field Test
DBF	Digital Beam Forming
DIAL	Differential Absorption Lidar
DMSP	Defense Meteorological Satellite Program
DOC	U.S. Department of Commerce
DoD	U.S. Department of Defense
DOE	U.S. Department of Energy
DOI	U.S. Department of Interior
DOT	U.S. Department of Transportation
DOW	Doppler on Wheels
DQA	Data Quality Analysis

DTASS	Digital Terminal Area Surveillance System
ELDORA	Electra Doppler Radar
EPA	U.S. Environmental Protection Agency
ERS-1 and 2	European Remote Sensing Satellite
FAA	Federal Aviation Administration
FEMA	Federal Emergency Management Administration
FHWA	Federal Highway Administration
FPGA	Field Programmable Gate Arrays
GPM	Global Precipitation Mission
GPS	Global Positioning System
IOS	Integrated Observing System
ITWS	Integrated Terminal Weather System
JPOLE	Joint Polarization Experiment
LNA	Low Noise Amplifier
MESAR	Multifunction Electronically Scanned Aperture Radar
MMIC	Monolithic Microwave Integrated Circuit
MOTR	Multiple Object Tracking Radar
MPR	Microburst Prediction Radar
NAS	National Airspace System
NASA	National Aeronautics and Space Administration
NCAR	National Center for Atmospheric Research
NCDC	National Climatic Data Center
NCEP	National Centers for Environmental Prediction
NEXRAD	Next Generation Weather Radar
NLDN	National Lightning Detection Network
NMOC	Naval Meteorological and Oceanography Command
NOAA	National Atmospheric and Oceanic Administration
NPI	NEXRAD Product Improvement Program
NPS	National Park Service
NRC	Nuclear Regulatory Commission
NSCAT	NASA Scatterometer
NWP	Numerical Weather Prediction
NWS	National Weather Service
OAO	Office of Aircraft Operations
ONR	Office of Naval Research
OPUP	U.S. Air Force Open Principal User Processor
ORDA	Open Radar Data Acquisition
ORPG	Open-System Radar Product Generator
PUP	Principalmanned User Processor
QPE	Quantified Precipitation Estimation
QPF	Qualitative Precipitation Forecasting
RASS	Radio Acoustic Sounding
RDA	Radar Data Acquisition

RF	Radio Frequency
ROC	Radar Operation Center
RPG	Radar Product Generator
SAR	Synthetic Aperture Radar
SSM/I	Special Sensor Microwave/Imager
T/R	Transmit/Receive
TASS	Terminal Area Surveillance System
TDWR	Terminal Doppler Weather Radar
TRMM	Tropical Rainfall Measurement Mission
TWT	Traveling Wave Tube
UAV	Unmanned Aerospace Vehicles
USAF	United States Air Force
USCG	U.S. Coast Guard
USDA	U.S. Department of Agriculture
USGS	U.S. Geological Survey
USWRP	U.S. Weather Research Program
VAD	Velocity Azimuth Display
VCP	Volume Control (Coverage) Pattern
WATADS	Algorithm Testing and Display System
WMO	World Meteorological Organization
WRC	World Radio Committee
WSDDM	Weather Support to Deicing Decision Making

Appendix C

Committee and Staff Biographies

Dr. Paul L. Smith (Chairman) is the past director of the Institute of Atmospheric Sciences of the South Dakota School of Mines and Technology and is now a professor emeritus in the Institute. He received his Ph.D. in electrical engineering from Carnegie Institute of Technology (now Carnegie-Mellon University). He served as chief scientist at Air Weather Service (AWS) Headquarters, Scott Air Force Base, during 1974–1975 and received the Award for Meritorious Civilian Service for his contributions to the AWS radar program. He served on the Executive Committee of the International Commission on Clouds and Precipitation from 1988 to 1996, as director of the South Dakota Space Grant Consortium from 1991 to 1996, and as a member of the National Research Council's National Weather Service (NWS) Modernization Committee from 1997 to 1999. He currently serves on the NEXRAD Technical Advisory Committee. His major research interests are in radar meteorology, with emphasis on quantitative measurement techniques and physical interpretation of the data; cloud physics, with an emphasis on studies of storm microphysics and kinematics using aircraft and radar; and weather modification, with an emphasis on the design and evaluation of experimental and operational projects. He manages the armored T-28 research aircraft facility and has worked on the development of various types of meteorological instrumentation. Dr. Smith is a fellow of the American Meteorological Society (AMS), and he chaired the AMS Committee on Radar Meteorology on two separate occasions. He received the 1992 Editor's Award from the AMS *Journal of Applied Meteorology*. He is Life Senior Member of the Institute of Electrical and Electronics Engineers (IEEE), a member of the Weather Modification Association (receiving its 1995 Thunderbird Award) and a member of Sigma Xi.

Dr. David Atlas is a distinguished visiting scientist, NASA Goddard Space Flight Center, Greenbelt, Maryland, and a consulting meteorologist with Atlas Concepts, Bethesda, Maryland. His primary interests are in radar meteorology, precipitation physics, and remote sensing. Tangential interests and experience are in the areas of weather hazards to aviation and the use of air- and space-borne synthetic aperture radar (SAR) over the ocean. He has developed a wide variety of devices and techniques for use in radar detection and warning and has studied many atmospheric phenomena including all types of storms and clear air phenomena such as clear air turbulence and the boundary layer. Dr. Atlas has a broad familiarity with processes of precipitation growth in various kinds of storms and climatic regimes. In recent years he has focused much of his attention on the use of radar and other remote sensors for the quantitative measurement of rainfall from space and for ground truth of space-borne measurements. The aim of this work is to determine how rainfall in the tropical regions of the earth as influences the global circulation and relates to global climate change. Dr. Atlas has served on numerous National Research Council (NRC) committees, is a member of the National Academy of Engineering, and has recently been elected as an honorary member of the American Meteorological Society.

Dr. Howard B. Bluestein is professor of meteorology at the University of Oklahoma, where he has served since 1976. He received his Ph.D. in meteorology from the Massachusetts Institute of Technology. His research interests are the observation and physical understanding of weather phenomena on convective, mesoscale, and synoptic scales. Dr. Bluestein is a fellow of the American Meteorological Society (AMS) and the Cooperative Institute for Mesoscale Meteorological Studies. He is past chair of the National Science Foundation (NSF) Observing Facilities Advisory Panel, the AMS Committee on Severe Local Storms, and the University Corporation for Atmospheric Research's (UCAR) Scientific Program Evaluation Committee, and he is a past member of the AMS Board of Meteorological and Oceanographic Education in Universities. He is also the author of a textbook on synoptic-dynamic meteorology and of *Tornado Alley*, a book for the scientific layperson on severe thunderstorms and tornadoes.

Dr. V. Chandrasekar is a professor in the Electrical Engineering Department at Colorado State University, where he received his Ph.D. Dr. Chandrasekar has been involved with weather radar systems for over 20 years and has about 25 years of experience in radar systems. Dr. Chandrasekar has played a key role in developing the CSU-CHILL radar as one of the most advanced meteorological radar systems available for research. He is continuing to work actively with the CSU-CHILL radar, supporting its research and education mission. He specializes in developing new radar technologies and techniques for solving meteorological problems. He has actively pursued applications of polarimetry for cloud microphysical applications, as well as neural network based radar rainfall estimates and fuzzy logic systems for hydrometeor identification. He is an avid experimentalist, conducting experiments to collect in situ observations to verify the new techniques and technologies. He is coauthor of two textbooks: Polarimetric and Doppler Weather Radar (Cambridge University Press) and Probability and Random Processes (McGraw Hill).

Dr. Eugenia Kalnay is chair of the Department of Meteorology at the University of Maryland, College Park. Her interests are in predictability and ensemble forecasting, numerical weather prediction, and data assimilation. With her collaborator Dr. Zhao-Xia Pu, she has introduced the method of backward integration of atmospheric models and several novel applications such as FAST 4-D VAR, and targeted observations. With her collaborator Dr. Zoltan Toth, she introduced the breeding method for ensemble forecasting. She is the author (with Ross Hoffman and Wesley Ebisuzaki) of other widely used ensemble methods known as Lagged Averaged Forecasting and Scaled LAF. She has also published papers on atmospheric dynamics and convection, numerical methods, and the atmosphere of Venus. She was elected into the National Academy of Engineering for advances in understanding atmospheric dynamics, numerical modeling, and atmospheric predictability and in the quality of U.S. operational weather forecasts.

Dr. R. Jeffrey Keeler, a senior research engineer with the Remote Sensing Facility at the National Center on Atmospheric Research's Atmospheric Technology Division (NCAR/ATD), coordinates surface-based remote sensing and NEXRAD development activities plus advanced remote sensing techniques. Dr. Keeler's research has focused on signal processing techniques for remote sensors, including microwave weather radars, optical sensors (lidar), UHF wind profilers, and acoustic sounding systems. Currently he is developing adaptive optimization techniques for fuzzy logic networks. At NCAR, Dr. Keeler has been responsible for advanced radar techniques using phased array antennas and complex pulse compression waveforms. He has assisted the Federal Aviation Administration (FAA) in planning the next generation aircraft and weather surveillance radar using these contemporary technologies and presently leads the NCAR program aimed at data-quality enhancements to the NEXRAD radars for the National Weather Service. In the 1970s, Dr. Keeler developed acoustic sounding data acquisition and processing equipment and digital processing techniques for Doppler lidar systems, and he managed a large simulation model of a satellite wind-finding lidar for global wind measurements. Previously, at Bell Laboratories, he designed and built a working model of the first digital equalizer, which has become commonplace in today's PC modems. Dr. Keeler is an adjunct professor at Colorado State University, teaching portions of meteorological radar courses, and he has taught in-depth remote sensing courses within NCAR and at foreign weather services. Dr. Keeler received his Ph.D. in electrical engineering from the University of Colorado.

Dr. John McCarthy is the manager of Scientific and Technical Program Development at the Naval Research Laboratory in Monterey, California. Previously, Dr. McCarthy served as Special Assistant for Program Development to the Director of the National Center for Atmospheric Research (NCAR) in Boulder, Colorado. Prior to that, he served as the director of the Research Applications Program (RAP) at NCAR. As director of RAP, he directed research associated with aviation weather hazards including NCAR activities associated with the Federal Aviation Administration (FAA) Aviation Weather Development Program, the FAA Terminal Doppler Weather Radar Program, and a national icing/winter storm research program. Previously, he directed NCAR activities associated with the Low-Level Windshear Alert System (LLWAS) project, which addressed the technical development of sensing systems to detect and warn of low-altitude windshear, the Joint Airport Weather Studies (JAWS), and the Classify, Locate and Avoid Wind Shear (CLAWS) project at NCAR. Additionally, Dr. McCarthy was the principal meteorologist associated with the development of the FAA Wind Shear Training Aid. In January 2000, Dr. McCarthy was named a fellow of the American Meteorological Society. Since the beginning of his tenure at NRL, Dr. McCarthy has developed programs in improving ceiling and visibility forecasting and in flight operations risk assessment, and he developed a broad program effort to improve short-term weather information to Navy battlegroups, entitled "NOWCAST for the Next Generation Navy." Dr. McCarthy received his Ph.D. in geophysical sciences from the University of Chicago.

Dr. Steven A. Rutledge heads the Department of Atmospheric Science at Colorado State University. His interests include mesoscale meteorology, atmospheric electricity, radar meteorology, and cloud physics. Dr. Rutledge is a member of the NASA TRMM Science Team. He was a cochair for the MCTEX Experiment and the STERAO-A Deep Convection Experiment. He was also the lead scientist for the shipboard radar program in TOGA COARE. He was the lead scientist for the TRMM-LBA Field Campaign in Brazil in January and February 1999. He also serves as the scientific director of the CSU-CHILL National Radar Facility, overseeing all engineering and scientific activities. In 1995 he served as chair of the 27th American Meteorological Society (AMS) Conference on Radar Meteorology. He is a member of the National Science Foundation's Facilities Advisory Council, which allocates aircraft and radar resources for field projects sponsored by the National Science Foundation (NSF). He is a past member of the AMS Committees on Radar Meteorology and Cloud Physics. Presently he serves as an associate editor for the *Journal of Geophysical Research*; he has also served as editor for *Geophysical Research Letters* and as an associate editor for *Monthly Weather Review*.

Dr. Thomas A. Seliga is an electronics engineer in the Surveillance and Sensors Division of the Volpe National Transportation Systems Center in the U.S. Department of Transportation. Dr. Seliga received his Ph.D. in electrical engineering from Pennsylvania State University. Dr. Seliga's research interests have been primarily in topics related to electromagnetic wave propagation and scattering, focusing mostly on problems in radio propagation and radar meteorology; he has also studied air pollution dispersion modeling and rainfall scavenging of aerosols. In the area of radar remote sensing, he originated and furthered the development of the concepts of differential reflectivity ZDR and specific differential phase shift KDP. These and other related polarimetric measurements have revolutionized the field of radar meteorology by improving radar's ability to quantify rainfall rates, detect hail, and discriminate between water and ice phase hydrometeors, and they have advanced the understanding of cloud physics. Most recently, he recognized and demonstrated the value of applying measurements from the NEXRAD radar system and other elements of the National Weather Service's Modernization and Associated Restructuring Program to surface transportation. He is a member of the American Association for the Advancement of Science, the American Geophysical Union, the Institute of Electrical and Electronics Engineers (Fellow), the American Meteorological Society, and the American Society for Engineering Education.

Dr. Robert J. Serafin, former director of the National Center for Atmospheric Research (NCAR) in Boulder, Colorado, was President of the American Meteorological Society (AMS) for 2001. He holds a Ph.D. in radar meteorology from the Illinois Institute of Technology. In 1989 he was appointed director of NCAR. The holder of three patents, Dr. Serafin has published more than 50 technical and scientific papers. He established the *Journal of Atmospheric and Oceanic Technology* and was its coeditor for several years. He has served on several National Research Council (NRC) panels and committees and was chair of the NRC Committee on National Weather Service Modernization. He also chairs a committee that advises the Federal Aviation Administration, the U.S. Air Force, and the National Weather Service on the nation's Doppler weather radar system. He is a member of the National Academy of Engineering, a fellow of the AMS, and a fellow of the Institute of Electrical and Electronics Engineers.

Dr. F. Wesley Wilson, Jr. is currently with National Center for Atmospheric Research's Research Applications Program (NCAR/RAP) and recently retired from Massachusetts Institute of Technology's (MIT) Lincoln Laboratory, where he led the Federal Aviation Administration's (FAA) Ceiling and Visibility Product Development Team and the first ceiling and visibility field project, the San Francisco Marine Stratus Initiative. Dr. Wilson received his Ph.D. from the University of Maryland, College Park. Through his career, Dr. Wilson's research emphasis has shifted from pure mathematics (differential topology and dynamical systems) to numerical and applied mathematics, and now to the current emphasis on the development of prototype operational meteorological systems and statistical forecast models. Dr. Wilson's previous contributions to FAA weather products include the development of two LLWAS algorithms, development of a portion of the TDWR microburst algorithm, and development of the ITWS gridded winds algorithm.

Dr. Vaughan C. Turekian is a program officer in the National Academy of Science's Board on Atmospheric Sciences and Climate and the Program Director for the Committee on Global Change Research. He received his B.S. degree from Yale University in geology and geophysics and international studies and his Ph.D. in environmental sciences from the University of Virginia. Dr. Turekian has been study director for a number of NAS studies including the recent Climate Change Science report requested by the White House.